Moving a Design into Production

Practical Engineering Guides for Managing Risk

by AT&T and the Department of the Navy

Design to Reduce Technical Risk
Design's Impact on Logistics
Testing to Verify Design and Manufacturing Readiness

Moving a Design into Production

AT&T

McGraw-Hill, Inc.

New York   San Francisco   Washington, D.C.   Auckland   Bogotá
Caracas   Lisbon   London   Madrid   Mexico City   Milan
Montreal   New Delhi   San Juan   Singapore
Sydney   Tokyo   Toronto

Library of Congress Cataloging-in-Publication Data

Moving a design into production / AT&T.
     p.   cm.
    Includes bibliographical references.
    ISBN 0-07-002564-9
    1. Production planning. I. American Telephone and Telegraph
Company.
    TS176.M69   1993
    658.5—dc20   92-12233
                                                CIP

1 2 3 4 5 6 7 8 9 0 DOC/DOC 9 9 8 7 6 5 4 3

ISBN 0-07-002564-9

The sponsoring editor for this book was Robert W. Hauserman, the editing supervisor was Kimberly A. Goff, and the production supervisor was Pamela A. Pelton. This book was set in Century Schoolbook. It was composed by McGraw-Hill's Professional Book Group composition unit.

Printed and bound by R. R. Donnelley & Sons Company.

Contents

How This Book Is Organized

This book has five major sections:

Introduction. Contains template definitions, explains which templates are covered, gives risks for each template, and states briefly how to reduce the risks.

Procedures. Presents the steps for reducing the risks for each template covered in this book.

Application. Gives examples for the principles and procedures discussed in the Procedures chapter.

Summary. Outlines information from earlier chapters.

References. Gives an annotated bibliography of sources for more information.

Acknowledgments

The following individuals and their respective organizations are hereby recognized for their contributions to the development of the texts on Practical Engineering Guides for Managing Risk.

Bell Labs, Holmdel, New Jersey

Dr. Robert D. Lake

Dr. Margaret Judith Doran

George J. Hudak

David A. Brittman

Ann D. Wright

Gus de los Reyes

Upendra Chivukcula

Dr. Blake Patterson

Dr. Howard H. Helms

Dr. Behrooz Khorramian

Julie Strachie

David B. Demyan

Valerie Mehlig

Federal Systems, Greensboro, North Carolina

James H. Everett

Robert S. Doar

David L. Hall

Teresa B. Tucker

Clydy D. Gann

Doug H. Weeks

Russell M. Pennington

J. Gil Jesso

Albert T. Mankowski

Office of Assistant Secretary of the Navy (Research, Development & Acquisition) Product Integrity Directorate, Washington, D.C.

Willis J. Willoughby, Jr.

Douglas O. Patterson

Edward L. Smith, Jr.

George E. Maccubbin

Louis C. Gills

Joseph G. Cady

General Electric, Arlington, Virginia

William A. Finn

Elwood P. Padgett, Jr.

Strategic Manufacturing Planning

1

Introduction

To the Reader

This part discusses five templates: Manufacturing Strategy, Manufacturing Plan, Tool Planning, Factory Improvements, and Qualify Manufacturing Process.

The templates, which reflect engineering fundamentals as well as industry and government experience, were first proposed in the early 1980s by a Defense Science Board task force of industry and government leaders, chaired by Willis J. Willoughby, Jr. The task force sought to improve the effectiveness of the transition from development to production of systems. The task force concluded that most program failures were due to a lack of understanding of the engineering and manufacturing disciplines used in the acquisition process. The task force then focused on identifying engineering processes and control methods that minimize technical risks in both government and industry. It defined these critical events in design, test, and production in terms of templates.

The template methodology and documents

A template specifies:

- areas of technical risk
- fundamental engineering principles and proven procedures to reduce the technical risks

Like classical mechanical templates, these templates identify critical measures and standards. By using the templates, developers are more likely to follow engineering disciplines.

In 1985, the task force published 47 templates in the DoD *Transition from Development to Production* (DoD 4245.7-M) manual.[1] The templates cover design, test, production, management, facilities, and logistics.

In 1986, the Department of the Navy issued the *Best Practices* (NAVSO P-6071) manual,[2] which illuminates DoD practices that increase risks. For each template, the *Best Practices* manual describes:

- potential traps and practices that increase the technical risks
- consequences of failing to reduce the technical risks
- an overview of best practices to reduce the technical risks

The *Best Practices* manual seeks to make practitioners more aware of traps and pitfalls so they can avoid repeating mistakes.

In September 1987, the Army Materiel Command made the templates the foundation for their risk reduction roadmaps for program managers.[3] In February 1991, the templates were incorporated into the DoD 5000.2 document as part of core of fundamental policies and procedures for acquisition programs.[4]

The templates are the foundation for current educational efforts

In 1988, the government initiated an educational program, "Templates: Professionalizing the Acquisition Work Force." This program includes courses and reference guides, such as this one, that increase awareness and promote the use of good engineering practices.

The key to improving the DoD's acquisition process is recognizing that the process is an industrial process, not an administrative process. This is a change in perspective that implies a change in the skills and technical knowledge of the acquisition work force in government and industry. Many in this work force do not have engineering backgrounds. Those with engineering backgrounds often do not have broad experience in design, test, or production. The work force must understand

---

[1]Department of Defense. *Transition from Development to Production.* DoD 4245.7-M, September 1985.

[2]Department of the Navy. *Best Practices: How to Avoid Surprises in the World's Most Complicated Technical Process.* NAVSO P-6071, March 1986.

[3]U. S. Army Materiel Command. *Program Management Risk Reduction Roadmaps.* Alexandria, VA: U.S. Army Materiel Command, September 1987.

[4]Department of Defense. *DoD Instruction 5000.2. Defense Acquisition Management Policies and Procedures.* Washington, DC: Department of Defense, February 23, 1991.

basic design, test, and production processes and associated technical risks. The basis for this understanding should be the templates which highlight the critical areas of technical risk.

The template educational program meets the needs of the acquisition work force. The program consists of a series of courses and technical books. The book provides background information for the templates. Each book covers one or more closely related templates.

How the parts relate to the templates. The parts describe:

- the templates, within the context of the overall acquisition process
- risks for each included template
- best commercial practices currently used to reduce the risks
- examples of how these best practices are applied

These books do not discuss government regulations, standards, and specifications, because these topics are well-covered in other documents and courses. Instead, the books stress the technical disciplines and processes required for success.

Clustering several templates in one reference guide makes sense when their best practices are closely related. For example, the best practices for the templates in Part 2, Parts Selection and Defect Control, interrelate and occur iteratively within design and manufacturing. Designers, suppliers, and manufacturers all have important roles. Other templates, such as Design Reviews, relate to many other templates and thus are best dealt with in other books.

Courses on the templates. The books are designed to be used either in courses or as stand-alone documents. An introductory course on the templates and several technical courses will be available soon.

The courses will help government and industry program managers understand the templates and their underlying engineering disciplines. The managers should recognize that adherence to engineering discipline is more critical to reducing technical risk than strict adherence to government military standards. They should especially recognize when their actions (or inactions) increase technical risks as well as when their actions reduce technical risks.

The templates are models

The templates defined in DoD 4245.7-M are not the final word on disciplined engineering practices or reducing technical risks. Instead, the templates are references and models that engineers and managers can apply to their industrial processes. Companies should look for high-risk technical areas by examining their past projects, by consulting their experienced engineers, and by considering industry-wide issues. The result of these efforts should be a list of areas of technical risk and best practices which becomes the company's own version of the DoD 4245.7-M and NAVSO P-6071 documents. Companies should tailor the best practices and engineering principles described in the reference guides to suit their particular needs. Several military suppliers have already produced manuals tailored to their processes.

Figure 1 shows where to find more and more details about risks, best practices, and engineering principles. Participants in the acquisition process should refer to these resources.

A major milestone in any product development process is the production of the first working model or the system itself. The processes by which products are developed and produced from one product to another are similar and apply to products from both the commercial and DoD sectors.

The development process starts with the perceived need for a product that performs a currently unavailable function, or the identification of an exploitable technological opportunity. Commercial companies continually evaluate their market segments for new product opportunities. Similarly, DoD continuously reviews the operational missions

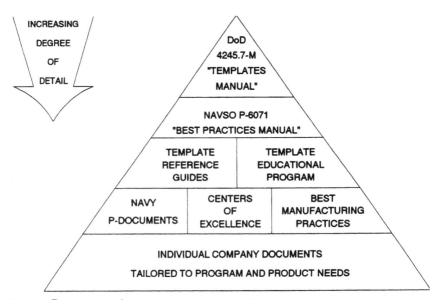

Figure 1 Resources on the acquisition process templates.

assigned to its forces to determine areas that are not adequately served by the available assets. Critical to the timely deployment of any system is manufacturing.

Manufacturing Planning Issues

The issues related to manufacturing planning are complex. Manufacturing planning involves many engineering disciplines, including design, producibility, manufacturing, quality, reliability, tooling, testing, and process engineering. It requires the step-by-step determination and verification of processes (e.g. tooling) and procedures (e.g. quality control) in the manufacture of a product. The planning efforts must also include the equipment, subcontractors, and suppliers. Manufacturing planning must be done properly and in a timely manner to ensure high-quality production.

Inadequate manufacturing planning can increase program risk. Understanding customer expectations facilitates the proper financial and personnel-resource decisions necessary for modernization, subcontractor structure, and technology-sharing agreements.

Furthermore, concurrent engineering philosophy stresses the importance of an integrated approach to product and process design. *Design to Reduce Technical Risk* reinforces the idea that manufacturing is not an afterthought to design.

Manufacturing Planning and the Acquisition Process

Contractors are not the only ones who must share in the responsibility of manufacturing planning. In a recent report requested by the US Army Materiel Command, the Manufacturing Studies Board (a part of the National Academy of Sciences) determined that the DoD's traditional acquisition process makes it difficult for contractors to justify the long-term investments required for the effective implementation of advanced manufacturing technologies. The initial sentence of the executive summary reads: "The gap between military requirements and industrial capabilities is in danger of becoming unmanageable in a national emergency, particularly conventional war, if the Department of Defense continues to neglect industrial preparedness planning."[5]

---

[5]Barlas, Stephen, "Army Takes Flak Over Flexible Manufacturing," *Managing Automation,* December 1990, 5(12), p. 20.

A report issued by the US Congress' Office of Technology Assessment suggested several ways the government can encourage improvements in US manufacturing competitiveness. Among the ideas were such things as lowering the cost of capital for investment, diffusing technology to smaller companies, taking increased responsibility for education and training, and sharing industry's costs for the development of strategic technologies.[6]

The government and industry must learn to support each other, especially in the defense community, if US manufacturing is to stay competitive. Industry, especially management, must take the lead to prevent problems before they occur, rather than waiting for a failure to occur before fixing it.

Background

Intense competition in the world market is forcing US manufacturers to rethink the way they design and produce parts. The concerns—time to market, quality, and cost—are putting pressure on equipment suppliers to develop machines that produce more and better parts in a shorter time, and that can be altered quickly to meet changing market demands.

Challenges facing the defense community

The top 25 defense contractors in the United States account for approximately half of the total contract dollars awarded by the Department of Defense (DoD). It is interesting to note that of the top 25 contractors, the DoD contracts ranged from about 4% to almost 100% of a company's revenues.[7]

Defense contractors have begun to realize that diversification into commercial markets can serve to provide revenues and stabilize their work force. The recent emphasis on a reduced defense budget has also served to reinforce this point to the defense community. Innovative approaches to manufacturing are required to meet this need.

Additionally, the US defense community is not insulated from the pressures of foreign competition. Many programs are often jointly competed from within, or are developed in, such organizations as NATO. Programs such as the Foreign Comparative Testing (FCT) program, assess foreign equipment to fill specific requirements. To be considered

---

[6]Gorte, Julie Fox, "Competing in Manufacturing: What Industry Can Do," *Manufacturing Systems,* January 1991, 9(1), pp. 63-64.

[7]Hammes, Sara and Tricia Welsh, "The Top 25 Contractors," *Fortune,* 123(4), February 25, 1991, pp. 68-69.

for FCT, a system or item must offer major performance, cost, or schedule advantages over similar US systems. The FCT program allows the military to purchase items off-the-shelf at a lower cost than if developed and produced the system itself. Some examples of systems acquired as a result of FCT include a German made vehicle for nuclear, biological, and chemical reconnaissance, which was deployed during Operations Desert Shield and Desert Storm.[8]

How manufacturing is perceived

Unfortunately, many hurdles prevent an easy transition from design to manufacturing. Historically, the manufacturing was perceived as "unglamorous," or even tedious. As a result, the perception of manufacturing people as having "dirty fingernails" has built a wall between design and manufacturing.

Manufacturing is seen as something you do when you are not designing.[9]

Manufacturing engineers often complain about lack of respect. Often they are viewed as a "necessary evil" to bridge the gap between design and production.

In a study conducted by A.T. Kearney for the Society of Manufacturing Engineers, CEOs were asked to assess manufacturing engineers. Less than half of the CEOs rated them as "good to excellent."[10] Lack of understanding of the manufacturing engineer's role may be the cause. Few people realize the number of "hats" manufacturing engineers must wear and the effort required on their part to successfully integrate all the tasks they face.

Manufacturing as a competitive advantage

Although the concept of manufacturing as a strategic weapon is not new, it is gaining a new level of respect. Over the past few decades, manufacturing's competitive thrust has evolved from one of product cost to one of *time.* Table 1[11] illustrates this evolution.

---

[8]Starr, Barbara, "Foreign Comparative Testing: Looking for the Best," *Jane's Defence Weekly,* 15(4), January 26, 1991, pp. 117-122.

[9]Koska, Detlef K. and Romano, Joseph D., *Countdown to the Future: The Manufacturing Engineer in the 21st Century,* Profile 21 Executive Summary, Dearborn, MI: SME, 1988, p. 14.

[10]Koska, Detlef K. and Romano, Joseph D., *Countdown to the Future: The Manufacturing Engineer in the 21st Century,* Profile 21 Executive Summary, Dearborn, MI: SME, 1988, p. 14.

[11]Adapted from Koelsch, James R., "Manufacturing Strategy: The Secret Weapon," *Machine and Tool BLUE BOOK,* September 1989, p. 47.

TABLE 1   The Evolution of Strategic Manufacturing

	1960s	1970s	1980s	1990s
Competitive Thrust	Cost	Market	Quality	Time
Manufacturing Strategy	High Volume	Functional Integration	Process Control	New Product Introduction
	Cost Minimization	Closed Loop	Material Velocity	Responsiveness
	Stabilize		World Class Manufacturing	Manufacturing Metrics
	Product Focus		Overhead Cost	New Organization Forms
Manufacturing Systems	Production & Inventory Control System (PICS)	Material Requirements Planning (MRP)	Manufacturing Resource Planning (MRP II)	Computer Integrated Manufacturing (CIM)
	Numerical Control (NC)	Master Production Scheduling (MPS)	Just In Time (JIT)	Decentralization
		Computer Numerical Control (CNC)	Total Quality Control (TQC)	Simplification
			Computer-Aided Design & Manufacturing (CAD/CAM)	Total Quality Management (TQM)
				Self-Directed Work Force
				Activity-Based Costing

Strategic Planning and Management

Proper integration of the strategic, planning, and implementation issues associated with manufacturing requires a clear understanding of manufacturing's role, as well as a long-term perspective. Without a well-defined manufacturing strategy, companies may look for short-term solutions that may prove detrimental in the long run. Too often, strategic planning is treated lightly. However, it can be a powerful advantage in a competitive marketplace, and provide a sound foundation for doing business.

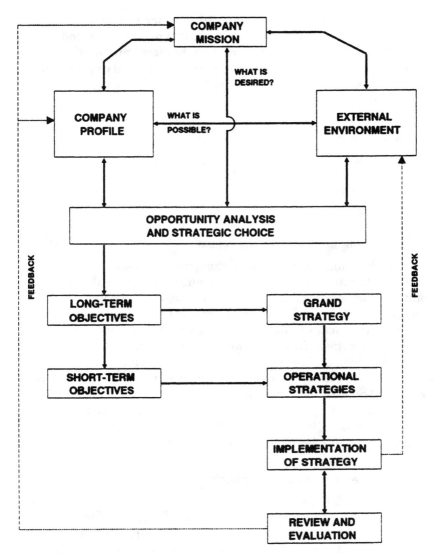

Figure 2 Strategic planning and management process.

Strategic planning and management is a dynamic, iterative, and *living* process. Through early establishment and proper maintenance, it can be a method for making sound strategic decisions at all levels of business operations. Figure 2 shows the strategic planning and management process.[12]

---

[12] Pearce and Robinson, p. 61.

The model provides context and direction for a process that relies heavily on communication and feedback. It is important to understand the objective of each block in Figure 2, as the blocks are tightly coupled. The entire process risks failure if one block is skipped or poorly implemented. In order to better understand the process, a summary of the model follows.[13]

Company mission

The key to the strategic management process is defining the *company mission,* which identifies the product, market, and technological areas for the business that are consistent with the values and priorities of the company.

Company profile and external environment

After establishing the mission the company assesses its internal and external environments. The *company profile* describes the company's abilities based on existing or attainable resources. The profile assesses inherent strengths and weaknesses of a company's structure. It also contrasts historical capabilities with current capabilities to identify future capabilities. *External environment* describes all conditions beyond the company's control, such as the economic, political, social, and technical forces at work in the company's general competitive environment. The external environment may also focus on a specific competitive situation external to the company that impacts strategic action. Generally, it describes actions taken by competitors, consumers, users, suppliers, creditors, or regulatory groups.

Opportunity analysis and strategic choice

The process of defining the mission and environments helps the company define the corporate and business unit directions. The first step is *opportunity analysis and strategic choice.* This process provides the long-term objectives and grand strategies that can position the company in the external environment to achieve its mission. Simultaneous assessment of the external environment and the company profile identifies opportunities that are possible areas of action. The company compares the list of opportunities against the company's mission to determine a set of possible and desirable opportunities.

---

[13]Pearce and Robinson, pp. 60-75.

Long-term objectives

Examining the list of opportunities enables the company to define *long-term objectives,* which state what is expected from a set of actions. Strategic manufacturing objectives should be specific, measurable, achievable, and consistent with other objectives within the company. Typically, strategic manufacturing objectives are defined by some or all of the following: profitability, return on investment, competitive position, technological leadership, productivity, employee relations, public responsibility, and employee development. The period for achieving these objectives is arbitrary, but five years is typical.

Grand strategy

To achieve the long-term objectives, the company defines a *grand strategy* which details a general plan of major actions to follow. Its purpose is to guide the acquisition and allocation of resources. While most grand strategies are unique, Pearce and Robinson identified 12 common approaches: concentration on the current business, market development, product development, innovation, horizontal integration, vertical integration, joint venture, concentric diversification, conglomerate diversification, retrenchment/turnaround, divestiture, and liquidation. Often, a grand strategy is a combination of several of these.

Short-term objectives

After identifying a set of long-term objectives and the corresponding strategies, management can begin to focus on the *short-term objectives.* Short-term objectives are more focused than the long-term objectives. They guide the short-term strategies and should reflect planning functions of the company's major divisions.

Operational strategies

Within each division or business unit is a set of *operational strategies,* which are specific and integrative plans of action for each distinctive function or division. Examples of operating or functional strategies are found in marketing, R&D, manufacturing, operations, and personnel.

Implementation of strategy

So far, most of the activities have been concentrated on defining objectives and goals. Now the *implementation of strategy* can begin. These are the actual activities of acquiring and allocating the resources and assigning responsibilities necessary to implement a strategy. Critical areas in a successful implementation are tasks, people, structures, technologies, and reward systems.

Review and evaluation

Critical to the successful implementation of the strategic management process is *review and evaluation*. The process is a dynamic and living thing; it is not a task that is performed once. Feedback is critical for correcting and maintaining direction. The ultimate test of a strategy is its ability to achieve its short-term objectives, long-term objectives, and mission. Strategies can only truly be evaluated after implementation. Therefore, it is critical to monitor and control the strategic plan.

Structure of This Part

The manufacturing planning process fits directly into the strategic management process shown in Figure 2. A manufacturing strategy is an *operational strategy*. The planning process is subordinate to grand strategy and is driven by the corporate mission and objectives. The subsequent planning and implementation activities are directed by the strategic activities that provide feedback in satisfying the company mission.

Five templates relevant to this part

Of the 48 templates identified in DoD 4245.7-M, five are germane to the proper handling of the material in this part:

Manufacturing strategy. Describes the overall, long-term framework that guides manufacturing functions. It is process oriented, consistent with an integrated approach of product and process design, and requires top-level corporate support.

Manufacturing plan. Details the actual approach and methods to follow. Identifies and addresses design, cost, scheduling, and tooling issues.

Factory improvements. Describes the need for state-of-the-art facilities and equipment. Describes how to evaluate and justify the degree of automation required.

Tool planning. Describes a process by which the equipment required for successful manufacture and test of the end product is identified. This process runs concurrently with process design and evaluates the make-or-buy decision for production tooling.

Qualify manufacturing process. Describes how to completely characterize the manufacturing process's capabilities. Also discusses the governments' Qualified Manufacturer's List (QML) efforts.

Strategic Manufacturing Planning

Strategic manufacturing planning refers to the concept of linking manufacturing to strategic planning and management, with the overall goal of making manufacturing a competitive advantage.

Strategic manufacturing planning takes on three different thrusts over the course of a program's life cycle. The three aspects can be described in terms of *strategic, planning,* and *implementation* activities. They occur in a chronological fashion but are tightly interconnected.

The strategic activities emphasize a clear vision of what the objectives are, while ensuring the issues of what is necessary to be successful. The planning activities address what is available to meet strategic objectives, while maintaining a realistic view of what is possible. The implementation activities represent the actions, tools, and techniques that must be identified and put in place in time for a smooth transition from strategy to plans to action.

The Procedures chapter of this part presents four major steps which span the set of strategic, planning, and implementation activities:

Develop Manufacturing Strategy relates the importance of manufacturing as a critical item in the way a firm conducts business.

Assess Present Capabilities and Desired Improvements discusses how, once an opportunity has been identified, a firm must examine its strengths and weaknesses in order to decide whether to pursue the opportunity.

Develop Manufacturing Plan details the importance of the manufacturing planning and management activities, and how they are critical for successful production.

Monitor, Control, and Improve discusses the need for understanding the manufacturing process and making sure it is in control and guaranteeing that continuous improvement is part of the overall environment.

Figure 3 illustrates the four steps of the overall process and reflects their connectivity with strategic planning and management model.

Strategic and planning activities

Before a contract is awarded, many manufacturing planning activities are directed towards answering an opportunity and making manufac-

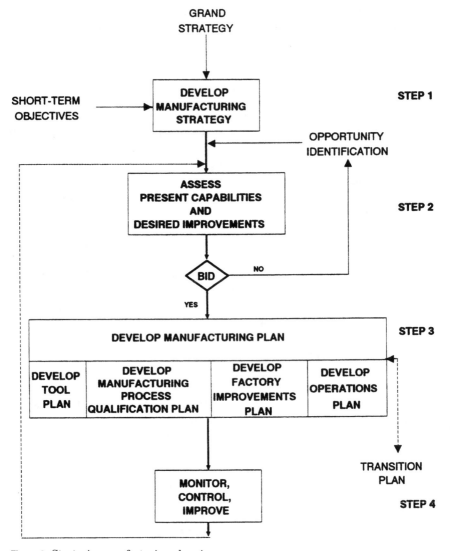

Figure 3 Strategic manufacturing planning.

turing a competitive weapon. These activities are reflected in Figure 4. During these activities, the manufacturing plan feeds the transition plan, helping program management identify potential risks and how to mitigate them.

Figure 4 represents the strategic activities combined with the early phases of the planning activities.

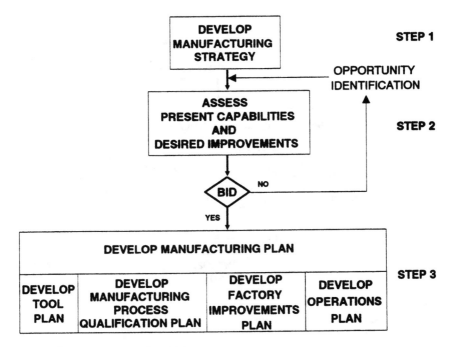

Figure 4 Pre-contract award activities.

Planning and implementation activities

After a contract has been awarded, activities are geared towards manufacturing planning to produce the system and towards continuous process improvement. This is reflected in Figure 5. Here, the manufacturing plan, as before, feeds the transition plan. However, the transition plan also feeds the manufacturing plan. This connectivity is critical for configuration management and program planning.

Figure 5 emphasizes how the planning activities are critical to the successful transition to implementation.

Risks

To remain competitive, manufacturing must continuously improve productivity and quality while reducing inventory costs and lead times. A corporate strategy for developing superior manufacturing capabilities is a major step toward increased productivity and quality. Unfortunately, large capital investment alone cannot immediately correct problems caused by years of neglect. Improving a company's manufacturing capa-

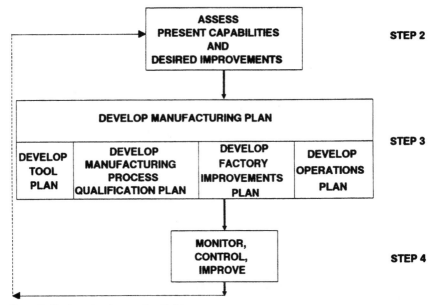

Figure 5 Post-contract award activities.

bilities is a long-term process that requires considerable reserves of both expertise and capital.[14] Risks exist in the strategic, planning, and implementation activities. Often, risks flow across each stage, and the consequence of one risk may become a risk to a later stage.

Strategic risks

Strategic decisions set the framework for long-term planning and action. Considering the full range of possible solutions can prevent strategic alternatives from being overlooked. A lack of understanding and awareness will result in short-sighted decisions or solutions that fail to address the real set of problems at hand.

A framework for strategic analysis is an attempt to summarize the entire range of activities involved in determining alternatives for the strategic allocation of resources.[15]

---

[14]Priest, J. W. *Engineering Design for Producibility and Reliability.* New York: Marcel Dekker, 1988, p. 51.

[15]D. R. Zeimer and P. D. Maycock, "A Framework for Strategic Analysis," in *Corporate Strategy and Product Innovation,* Robert R. Rothberg Ed., Free Press: NY, 1976, p. 87.

TABLE 2   Strategic Risks and Consequences

Risk	Consequences
No strategic planning	Strategies not formulated
	Difficulty in obtaining management support
	No attempt at continuous improvement
	No consideration to make-or-buy decisions or contract manufacturing
	Alliances, partnerships, or acquisitions not considered
	Unrealistic expectations cause poor solutions
Manufacturing's role not understood	Manufacturing not integrated into the development process
	Manufacturing not considered important to corporate and business unit strategies
Training not considered important	Management does not understand impact of decisions on manufacturing
	Work force does not perform efficiently and effectively
	Line and staff workers do not understand impact of manufacturing on business decisions
Resources and capabilities not examined and understood	Poor allocation of personnel, facilities, and money
Facility needs not examined or understood	Necessary manufacturing processes not identified
	No evaluation of alternatives such as focused factories, flexible manufacturing, or QML
	Inventory and material handling needs not addressed until it is too late
	Type, space required, and location of site not identified in time

Table 2 summarizes strategic risks and consequences that can be avoided.

Planning risks

Ineffective planning hinders performance. Effective planning requires flexibility and consistency with strategic decisions. If linkages between strategic activities and other planning activities are not maintained, assumptions and decisions cannot be verified. Similarly, implementation activities need to provide feedback to identify and correct errors in planning, or to provide redirection in order to keep activities aligned with strategic objectives.

TABLE 3   Planning Risks and Consequences

Risk	Consequences
Manufacturing management efforts not in place early	No manufacturing planning, or planning starts too late
	No manufacturing plan developed
	Manufacturing risks not identified and mitigated
Manufacturing not tied to a concurrent engineering or integrated product development (IPD) effort	Manufacturing plan has no connectivity and feedback with transition plan
	Manufacturing plan has no connectivity and feedback with risk management plan
	Producibility not considered during design
	Ineffective manufacturing process engineering
	Inadequate flow-down of requirements, plans, policies, strategies, standards, etc. to sub-contractors
Reliance on technology as the only solution	Money is thrown at problems
	Focus on technology and initiative implementation
	Dependence on automation to solve problems
No consideration to factory improvements, special tools, special test equipment, process qualification, and process operations	Site is not adequate to help meet cost and schedule objectives
	Manufacturing process is inefficient and ineffective
	Cannot properly build and test the system

Table 3 summarizes planning risks and consequences that can be avoided.

Implementation risks

Poor preparation and planning leads to reactive decision making, whose implementation does not adequately reflect strategic objectives. Managers may not have the proper resources to make in-process corrections. Reality can be a difficult and expensive lesson, leading to a game of catch-up rather than improvement and innovation.

Table 4 summarizes implementation risks and consequences that can be avoided.

Keys to Success

Strategic manufacturing planning is not trivial. However, by implementing an effective strategic manufacturing planning process, companies can become more competitive within their markets.

TABLE 4   Implementation Risks and Consequences

Risk	Consequences
No management support	Resources not allocated
Lack of continuous improvement tools and methods	Process improvements happen only once, if at all
	Data collection absent or insufficient
	Process monitoring and control absent or insufficient
Design, manufacturing, and test function independently	Cost, quality, schedule, and performance requirements may not be met
No configuration management and change control	Wrong design may be released to manufacturing
	Different areas (e.g., design, test, manufacturing, logistics) all working with different sets of requirements
	Poor documentation

It should be noted that the principles and practices associated with a successful implementation of the strategic manufacturing process are not unique to any industry. Manufacturing, when viewed as a competitive weapon, provides a good base for responsiveness and flexibility.

Implement strategic planning and management

Strategic planning and management is a process defined as "the set of decisions and actions resulting in the formulation and implementation of strategies designed to achieve the objectives of the organization."[16] Manufacturing decisions impact the day-to-day operations of any organization that produces products. Strategic manufacturing planning must be integrated with the overall strategic management process if manufacturing is to help achieve objectives.

Decisions regarding such issues as resources, partnerships/alliances (both internal and external), education and training, manufacturing objectives, and finance are *strategic* issues. All these efforts must be coordinated and consistent with those of all others in an organization, regardless of their place in the strategic hierarchy.

Furthermore, clearly defined corporate missions, objectives, strategies, and plans must be documented and communicated to all levels of the company. This job must be carried out by top management, who also provide the leadership and direction for all to follow. Strategic

---

[16]Pearce and Robinson, p. 4.

planning and management is an iterative process requiring constant monitoring, evaluation and updating of plans and strategies.

Maintain consistency, connectivity, and communication across all functions

As mentioned, strategic manufacturing planning includes strategic, planning, and implementation activities. For a successful completion of the process, feedback is critical for making corrections and keeping the manufacturing objectives aligned with those of the business unit or corporation. Additionally, the results of these activities provide valuable information and lessons learned for future endeavors.

Maintain links with a formal, dynamic transition plan

An important item in any project is the existence of a transition plan. *Design to Reduce Technical Risk* discusses the risks and keys to success in developing a transition plan.

A transition plan is a roadmap for a program. Its main purpose is to integrate the requirements of the design, test, production, facilities, logistics, and management functions into a cohesive plan that minimizes a project's cost, schedule, and technical risks.

The transition plan provides the link between the manufacturing plan and all other program plans. In order to minimize risks, the manufacturing plan both receives input from and provides feedback to the transition plan.

Implement a concurrent engineering philosophy

A concurrent engineering philosophy stresses the integrated design of product and process throughout an entire product's life cycle.

Whether program development efforts are referred to as concurrent engineering or integrated product development, coordinating the design of product and process is critical. Design, test, and manufacturing functions do not occur discretely. If a product is to be designed, tested, and built to requirements, it requires a team effort, with each discipline sharing and communicating concerns and feedback.

In strategic manufacturing planning, items such as special tooling, special test equipment, process plans, and facility concerns require consideration as early as possible. Additionally, much of the tooling and test equipment needs to be designed concurrently with the product and manufacturing processes. It is important that team composition reflects these concerns.

If a concurrent engineering philosophy is to flourish, configuration management and change control must be present. Design revisions must be adequately documented and communicated so that everyone is working on the same set of requirements.

Implement continuous improvement

Changes in technology, processes, and customer needs demand constant attention if a company is to remain competitive and profitable. Continuous improvement stresses tools and methods to help identify areas of change, opportunities for improvement, and plans to implement improvement.

By taking a process-oriented approach to continuous improvement, it is possible to map out the process in some usable fashion (e.g. a flowchart). This exercise serves several functions: it forces a team to think about the way the process works, identifies missing pieces to the process, and identifies value-added and non-value-added functions. With this information in hand, a team can map out a plan for improving the process.

As the name suggests, continuous improvement is an ongoing task, requiring diligence, discipline, and patience.

Ensure proper data collection and data management

Continuous improvement requires collecting and using data to drive improvements, facilitate process control, and indicate areas of excessive variation. A well-defined process and a set of measures or metrics, when analyzed, can provide a statistical measure of a process' performance. Examples of the types of data that can be collected include part yield, number of defects, operator hours, part costs, and material flow.

Statistical analyses can provide valuable information regarding: variability in the processes, which processes need improvement, and root causes. With this information, it is possible to perform such tasks as a robust design on a process to reduce its variability.

Data collection and analysis is not a one-time affair. Management of data is essential if statistical analyses are to be of any value.

Emphasize training and education

In his book *Out of the Crisis,* W. Edwards Deming presents 14 points for management and industry, in general, to follow if they are to increase the quality and productivity of the people and processes they manage. Of those points, two specifically address the importance of education, training, and self-improvement.

It has been suggested that of the possibilities for improvement, management is responsible for 85%–94% of them.[17][18] Similarly, Juran has proposed the "Pareto Principle,"[19] which emphasizes the *vital few* things that must be addressed individually and the *trivial many* that can be addressed as a whole.

If management is not capable of distinguishing possibilities for improvement, or cannot identify the vital few, it is unlikely that they will have a good background for making sound strategic decisions. It is critical for *everyone* to understand the implications of their decisions and actions.

Education and training serve to provide all employees with the proper knowledge base for making informed decisions. As with continuous improvement, training is not a one-shot thing. Due to the dynamic nature of business and technology, all employees must be kept up to date and aware in order to remain effective.

---

[17]Scherkenbach, William W., *The Deming Route to Quality and Productivity: Road Maps and Roadblocks*, CEEPress Books: Washington, 1988, p. 101.

[18]Deming, W. Edward, *Out of the Crisis*, Massachusetts Institute of Technology: Cambridge, MA, p. 315.

[19]Juran, Joseph M., *Managerial Breakthrough*, McGraw-Hill: NY, 1964, p. 44.

Chapter

2

Procedures

This part outlines the development of manufacturing strategy, manufacturing plans, tooling plans, manufacturing-process qualification plans, factory-improvements plans, and operations plans. This part also emphasizes monitoring and continuous improvement of various plans through the life cycle of a system.

Strategic manufacturing planning has three phases: strategic, planning, and implementation. Strategic manufacturing planning is the method to:

- develop a manufacturing strategy
- develop, implement, and maintain a manufacturing plan
- monitor and continuously improve the manufacturing strategy and plan

Developing a manufacturing strategy requires an integrated, focused, and strategic approach.[20]

Developing and implementing a manufacturing plan requires monitoring and feedback and is an iterative process. The manufacturing plan has to be a living document to achieve manufacturing goals.

This chapter presents procedures in strategic manufacturing planning in four steps as shown in Figure 6 (below).

Step 1 describes how to develop manufacturing strategy.

Step 2 describes how to assess present capabilities and desired improvements.

Step 3 describes how to develop a manufacturing plan.

Step 4 describes how to monitor, control, and improve the manufacturing plan.

---

[20]Skinner. "A Strategy for Competitive Manufacturing." *Management Review.* August 1987, p. 55

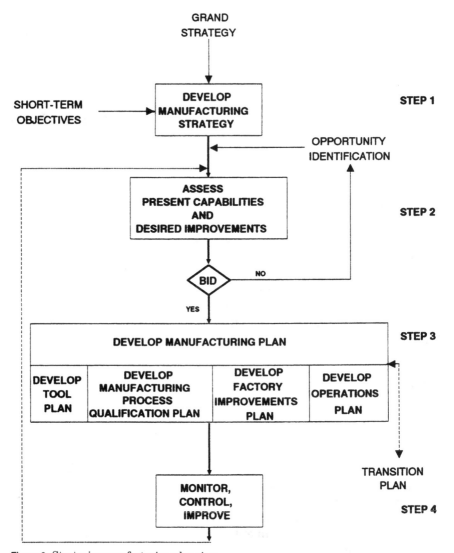

Figure 6 Strategic manufacturing planning.

Step 1—Develop Manufacturing Strategy

Manufacturing has entered the early stages of a revolutionary period caused by the convergence of three powerful trends:[21]

■ The rapid advancement and spread of manufacturing capabilities worldwide has created intense competition on a global scale.

---

[21]Manufacturing Studies Board. *Toward a New Era in U.S. Manufacturing: The need for a National Vision.* Washington, DC: National Academy Press, 1986, p. 1.

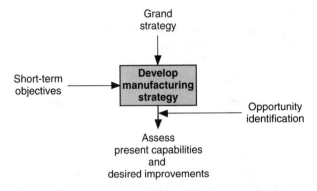

Emphasis on strategic and planning activities

Figure 7 Develop manufacturing strategy.

- The emergence of advanced manufacturing technologies is dramatically changing both the products and processes of modern manufacturing.

- Changes in traditional management and labor practices, organizational structures, and decision-making criteria represent new sources of competitiveness and introduce new strategic opportunities.

Devising business and manufacturing strategies will enable companies to address these trends, to manage and organize their businesses, remain competitive, and stay in business.

What is manufacturing strategy?

Webster's Dictionary defines strategy as:

The art of devising or employing plans toward a goal.

It is important to note that a strategy is future oriented and evolutionary. Strategies promote change over time, rather than an abrupt change, to achieve a goal or set of goals.

The Webster's Dictionary definition of manufacturing is:

To produce according to an organized plan and with division of labor.

Skinner has suggested that the *purpose* of manufacturing is "...to serve the company—to meet its needs for survival, profit, and growth."[22] A strategy that covers just manufacturing operations is restricting.

---

[22]Skinner, Wickham. "Manufacturing—Missing Link in Corporate Strategy." *Harvard Business Review.* May/June 1969, p. 140.

A manufacturing strategy is not just a piece of paper. It is a living, dynamic document that requires an understanding of what manufacturing is and how manufacturing fits into the overall operations of a business. A good definition of a manufacturing strategy is:

> ...a projected pattern of manufacturing choices formulated to improve fundamental manufacturing capabilities and to support a business and corporate strategy.[23]

Manufacturing strategy allows an organization to question where it is now, and where it must be for both the short term and the long term.

When formulating a strategy the following questions are often asked:[24]

- What is the goal of the business, and how does it get there?

- What does the company really make?

- How does the company go about making it?

- Where does the company want to make it?

- Who is going to make it?

- What information do the people who are going to make it need?

- What system provides the information?

Questions like these are part of an overall, iterative process of strategic thinking.

Role of manufacturing strategy

Successful companies often owe much of their success to well-developed manufacturing strategies. The manufacturing strategy establishes a long-term perspective for manufacturing. It is a critical, integral element of an overall business strategy—in manufacturing industries— that ensures that a company's manufacturing operations support the business strategy.

A corporate strategy comprises four strategies: financial, manufacturing, marketing, and product. For corporate strategy to be of value,

---

[23]Miller, Jeffrey G. and Hayslip, Warren. "Implementing Manufacturing Strategic Planning." *Planning Review*. July/August 1989, p. 23.

[24]Koelsch, James R. "Manufacturing Strategy: The Secret Weapon." *Machine and Tool BLUE BOOK*. September 1989, pp. 43-44.

Figure 8 Components of a corporate strategy.

Figure 8[25] suggests the coordination of the manufacturing strategy with other strategies.

There are several levels of strategy within the corporate structure. A business might have individual functional strategies such as marketing/sales, manufacturing, R&D, and accounting/control. A functional strategy specifies how it will support the desired business strategy and how it will complement other functional strategies. For each strategy to be effective, it must ensure consistent support of other strategies. Manufacturing strategy is an integral component of the overall corporate strategy.

Figure 9[26] illustrates the hierarchy among the different strategy levels: corporate, business, and functional.

Strategic decision areas

The driving force for developing the manufacturing strategy must come from the top. Without this support, meeting strategic objectives will be difficult.

Developing a manufacturing strategy requires addressing several decision areas. These manufacturing choices help identify and plan the functional nature of the manufacturing strategy.

---

[25]Koelsch, James R. "Manufacturing Strategy: The Secret Weapon." *Machine and Tool BLUE BOOK.* September 1989, p. 44.

[26]Wheelright, Steven C. "Manufacturing Strategy: Defining the Missing Link." *Strategic Management Journal,* vol. 5, 1984. p. 83.

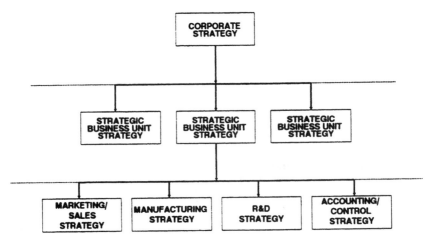

Figure 9 Levels of strategy.

The old paradigm of efficient manufacturing is "big is better." Departments are dedicated to a particular function. The belief is that if the machines are used fully, the people are kept busy, and the warehouse is full, the firm will show profits.

The new concept of "plant within a plant" or *focused factory* calls for increasing the demands and narrowing the scope of manufacturing operations. This can be achieved by concentrating management attention on a few key tasks and priorities.

A manufacturing strategy comprises a series of coordinated decisions. For example, decisions concerning facilities, capacity, technology, and vertical integration have long-term impacts, because reversing or undoing them once they are in place is difficult and may require substantial capital investment. Other decisions regarding such things as management policies and systems include work force, quality, production control, and organization. These decisions are more tactical, because they impact on-going decisions and their links to operations, and they do not require substantial capital investment.

Wheelright puts manufacturing choices into the eight categories shown in Table 5.[27] [28] [29]

---

[27]Wheelright, Steven C. "Manufacturing Strategy: Defining the Missing Link." *Strategic Management Journal,* vol. 5, 1984. p. 84.

[28]Wheelright, Steven C. and Hayes, Robert H. "Competing through Manufacturing." *Harvard Business Review.* January-February 1985, p. 100.

[29]Adapted from Pound, Ronald and Smith-Vargo, Linda. "Manufacturing Strategy Can Beat a Bum Rap." *Electronic Packaging & Production.* September 1988, p. 56.

TABLE 5   Strategic Manufacturing Decisions

Strategic Decision	Representative Issues
Capacity	Throughput (amount, timing, type), new purchases
Facilities	Factory (size, location, specialization)
Technology	Automation, fabrication methods
Vertical Integration	Make-or-buy (direction, extent, balance), in-house dedicated, contract assembly
Work force	Direct labor cost, skill levels, training
Quality	Definition, role, responsibility
Production Control	Work flow, Just-in-time, component sourcing, storage/retrieval, material handling
Organization	Integrating design and manufacturing

When developing a manufacturing strategy, management needs to recognize the alternatives and select appropriate alternatives consistent with the corporate strategy.

A guide to the understanding of each of these strategic decision areas is found in the Application chapter of this part.

Manufacturing's strategic role

Many companies strive to become what is commonly referred to as "world class." Historically, companies did not understand how manufacturing actually contributes towards the overall strategic goals of the company. Additionally, strategists often overlook the fact that transforming manufacturing from a corporate weakness to a competitive strength takes disciplined effort and top-level management support.

Another critical item is patience. Strategic planning implies an evolution over time towards a set of lasting goals. A corporate culture that promotes radical and revolutionary change, without considering the implications, tends to undermine the benefits of strategic planning. A "quick fix" most likely would result in short-term benefits and long-term disadvantages. It is important to understand how manufacturing must evolve into a competitive advantage. A wrong decision in such areas as equipment, location, and size can result in several years of problems. In order for manufacturing to complement business operations (as opposed to a detriment), companies require speed and flexibility.

Manufacturing effectiveness. Manufacturing does not just become a competitive advantage. Wheelright and Hayes have suggested that there are stages of manufacturing effectiveness:

TABLE 6   Stages in Manufacturing's Strategic Role

Stage	Characteristics
Stage 1—Minimize manufacturing's negative potential: "internally neutral"	Outside experts are called in to make strategic-manufacturing decisions.
	Internal, detailed management control systems are the primary means for monitoring manufacturing performance.
	Manufacturing is kept flexible and reactive.
Stage 2—Achieve parity with competitors: "externally neutral"	"Industry practice" is followed.
	The planning horizon for manufacturing-investment decisions is extended to incorporate a single business cycle.
	Capital investment is the primary means for catching up with competition or achieving a competitive edge.
Stage 3—Provide credible support to the business strategy: "internally supportive"	Manufacturing investments are screened for consistency with the business strategy
	A manufacturing strategy is formulated and pursued.
	Long-term manufacturing developments and trends are systematically addressed.
Stage 4—Pursue a manufacturing-based competitive advantage: "externally supportive"	Efforts to anticipate the potential of new manufacturing practices and technologies are made.

Table 6 summarizes the role of manufacturing in each of the four stages.[30]

When considering these stages, management must recognize a few key points:[31]

- The stages are not mutually exclusive. Manufacturing operations are often composed of several factors, each in a different stage of development. As a result, management must determine a level where all factors are balanced.

- It is difficult, if not impossible, to skip a stage. A manufacturing operation that is up and running has little freedom of choice. Established methods require a great deal of effort to change. It is important to master the activities at any level, as that often provides a foundation

---

[30] Adapted from Wheelright, Steven C. and Hayes, Robert H. "Competing through Manufacturing." *Harvard Business Review*. January-February 1985, p. 100.

[31] Wheelright, Steven C. and Hayes, Robert H. "Competing through Manufacturing." *Harvard Business Review*. January-February 1985, p. 100.

for a successful transition to the next stage. Otherwise, the less advanced activities tend to hold back the more advanced.

- The nuts-and-bolts of the development work takes place at the business unit level. While it may be appealing for an entire business entity to move through the stages, the critical coordination of functions takes place at the business unit. Support from the corporate level is important, but an environment that allows business units to evolve and help each other has a greater chance of success.

Companies that are in Stage 4 are generally those that are considered "world class." The changes required to reach Stage 4 involve fundamental changes in the perception and interaction of manufacturing with the rest of the organization.

Evaluate manufacturing choices

At this point in the process, the company should identify, understand, and address the different strategic manufacturing choices. Also, the company should be able to characterize manufacturing's strategic role. The next step in developing the manufacturing strategy is to evaluate all these choices. When evaluating manufacturing choices, two areas for action are:

- *Key capability development*—approaches the strategy from a learning and continuous improvement perspective, with emphasis on the choices that affect such things as: improving manufacturing's competitive capabilities, producing high-quality products, and rapid new product introduction.

- *Strategic-plan development*—focuses on establishing action plans, with emphasis on such things as: fundamental competitive abilities, linking manufacturing to product and market plans, and identifying resource constraints and opportunities.

To accomplish competitive advantage through manufacturing, the company must balance key capability development with the strategic plan development.

Figure 10[32] presents a matrix of possible outcomes during attempts to balance efforts between key capability development and strategic plan development.

For a manufacturing strategy to evolve, it must be given the chance from the start and the planning process must be allowed to improve as more experience is gained.

---

[32] Adapted from Miller, Jeffrey G. and Hayslip, Warren. "Implementing Manufacturing Strategic Planning." *Planning Review.* July/August 1989, pp. 24-25.

Key capability development		
	Strong	**Weak**
Strong	Manufacturing as a competitive advantage	Little potential, strong effort
Weak	Strong potential, wasted effort	Manufacturing as a competitive disadvantage

(Left axis label: Strategic plan development)

Figure 10 Balancing manufacturing strategy development.

Performance measurement

If the ultimate goal is manufacturing as a competitive advantage, strategic objectives, manufacturing choices, strategy development, and implementation all must complement each other. Measuring success (or failure) ensures that progress is moving towards the desired goals.

The purpose of manufacturing performance measures is to encourage manufacturing excellence. Manufacturing excellence refers to reduced variability and producing uniformly good products efficiently. It can be achieved by providing customers with a product that surpasses competition in value and reliability.

Manufacturing strategy drives manufacturing performance and must focus on competitive objectives. Performance measures can evolve as strategy evolves. They should provide perspective on manufacturing's direction and rate of improvement.

An effective manufacturing performance system is developed when it:[33][34]

---

[33]Touche Ross. *Operating Principles for the 1990s Phase 1: Assessment of the Elements Comprising World Class Manufacturing.* National Center for Manufacturing Sciences, Ann Arbor, MI, June 1989, pp. 30-48.

[34]McDougall, Duncan C. *Effective Manufacturing Performance Measurement Systems: How to Tell When You've Found One.* Boston University School of Management, February 1988.

- focuses on competitive variables—those the customer sees
- promotes learning
- serves to promote the success of the whole business
- demands improvement only on the strategically-chosen variables, while the others are not allowed to backslide

The two types of performance measures are: financial and operational (or non-financial) measures. Financial measures of performance are often expanded to cover three major areas:

- product costing
- target pricing
- investment justification

Operational measures of performance are often broken down in four major areas:

- productivity measurement, including labor and machinery
- measures of time-based competition, i.e., delivery, throughput, and flexibility
- quality measures
- competitive benchmarking

Evaluate manufacturing strategy

Once a manufacturing strategy is developed and understood, present capabilities assessed, and goals identified, it is important that our company evaluate the manufacturing strategy.

Identify opportunities to evaluate. The business strategy identifies which areas the company should pursue and the product demand in those areas.

A manufacturing strategy is usually developed with some idea of the intended market—lessons learned, previous successes, market forces, and competition.

As is often the case, an opportunity presents an idea for a product to address a need. The source for the idea can be anything from an improvement on an existing product, or the government asking the defense contractors for a new weapons system.

When an opportunity presents itself, a company must compare it with the manufacturing strategy, its strengths and weaknesses (both internal and external to the organization), and the market forces.

Establish evaluation criteria. Strategy development occurs through a decision-making process. This suggests that certain criteria may be used to evaluate the appropriateness of a given manufacturing strategy. Such criteria generally fall into two groups:

- consistency

  between manufacturing strategy and the overall business strategy

  between the manufacturing strategy and other functional strategies within the business

  among the decision types in the manufacturing strategy

  between the manufacturing strategy and the business environment, such as resources available, governmental restraints, etc.

- emphasis on competitive success factors

  making trade-offs explicit, allowing manufacturing to prioritize activities

  directing attention to opportunities that fit the business strategy

  promoting clarity regarding the manufacturing throughout the business unit

Implementing strategic manufacturing planning

To implement an effective manufacturing strategy, it is critical to establish and promote the proper environment. Miller and Hayslip have described five key points that are important in properly implementing strategic manufacturing planning:[35]

- **Understand why an integrated manufacturing strategy is important.** The outcome of a manufacturing strategy effort is not just a piece of paper or a bureaucratic exercise. It can produce a flexible and reactive organization. Additionally, it establishes priorities about what is important and why, ensuring effective implementation.

- **Obtain consensus on critical capabilities.** Consensus ensures that the key capability development and strategic planning activities are working towards the same goal.

- **Learn by doing—together.** Multifunctional teams provide a source of input and help develop a common language across organizations. Also, they help create a direct understanding of the manufacturing strategy process by getting the right people involved.

- **Plan how to get manufacturing up to corporate speed, and the corporation down to manufacturing fundamentals.** Education and training must be bi-directional. Technical people need to under-

---

[35]Miller, Jeffrey G., Hayslip, Warren. "Implementing Manufacturing Strategic Planning." *Planning Review.* July/August 1989, pp. 26-27, 48

stand the strategic planning process while nontechnical people (i.e., functional managers) must understand the impact of the planning process on manufacturing decisions. Strategic alliances also can help by providing a cooperative environment where ideas, experience, data, and even personnel can be freely exchanged.

- **Identify the appropriate manufacturing unit of analysis.** Careful consideration must be made in advance in order that the planning pieces fit together. A manufacturing unit does not always seem consistent with the notion of a strategic business unit. This is often because several products may be produced at any given location. By determining where manufacturing begins and ends, a company can achieve a greater understanding of how manufacturing fits into business operations.

The strategic manufacturing process is an evolutionary one. Too often, companies try to achieve strategic goals that are attempted by abrupt changes in, or implementations of initiatives within the company. In the alphabet soup of continuous improvement initiatives and philosophies, it is critical that management have a clear understanding of what each initiative is, intends to accomplish, and what it is **not.**

After the development of a manufacturing strategy and before the development of a manufacturing plan, the company needs to assess present capabilities and desired improvements.

Step 2—Assess Present Capabilities and Desired Improvements

Regardless of whether a firm is initially pursuing an opportunity, or looking for opportunities to continuously improve its manufacturing operations, a self-evaluation is warranted.

> Make no mistake, if you do not honestly recognize and quantify the present state and improvement tasks, ultimately the only one you might be kidding is yourself.[36]

Effective strategies for operations require a clear understanding of functions, activities, and tasks in producing the product. Before a company decides to embark on a project or program, it has to understand the program-specific needs and assess current capabilities.

Manufacturing planning should address the needs of a specific program and should be consistent with the business-unit and corporate plans.

---

[36]Subcommittee on Management Quality Change. *Staying Alive: Managing the Process of Change for Quality Improvement.* Dearborn, MI: American Supplier Institute, 1989, p. 11.

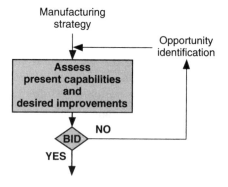

Emphasis on strategic and planning activities

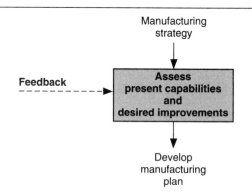

Emphasis on planning and implementation activities

Figure 11   Assess present capabilities and desired improvements.

Program-specific planning includes understanding the manufacturing needs of a given program. Planning should consider required resources, current capabilities, and whether the program is consistent with the strategic direction of the company. This planning helps the company decide whether to pursue an opportunity based on assessments of the company.

The basis for the assessments are a comparison of the "as is" condition of the business unit and the the "to be" conceptual plan alternatives. These comparisons are difficult and challenging tasks involving trade-offs, value judgments, educated guesses, and objective analyses.

Assess present capabilities

Assessing present capabilities identifies the strategically important strengths and weaknesses on which the company should plan and implement its strategy. In one study, managers were asked to evaluate

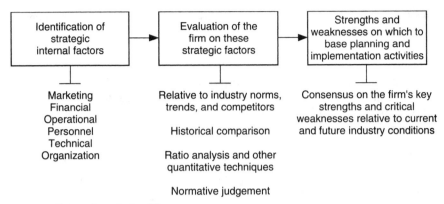

Figure 12 Internal analysis of the company.

apparent strengths on historical and relative-to-competitor bases. The results of the survey indicated that managers use different criteria for evaluating apparent strengths and potential weaknesses.[37] It is important that the evaluation be as balanced as possible, in order to reflect the true state of affairs.

Figure 12 illustrates the process of internal analysis.[38]

The assessment should focus on as many strategic areas as possible. Some important ones are: quality, product lead time, producibility, manufacturing processes, tooling, facilities, human resources, training capabilities, lessons learned, role in marketplace, strengths and weaknesses, performance measures, constraints, cost accounting, and make-or-buy. A brief discussion of each of these follows.

Quality. Current philosophy and procedures recognize that quality is not something that only results from the development or improvement of systems and equipment. Quality is also the result of overall attention to all activities throughout a system's life cycle.

The Malcolm Baldrige National Quality Award has helped motivate many US firms to adopt quality strategies. A key aspect of the Baldrige Award application is a self-assessment in many areas across the organization. Some Baldrige areas including leadership, strategic quality planning, quality results, customer satisfaction measures, and supplier involvement are critical to the success of strategic manufacturing.

---

[37]Stevenson, Howard H. "Defining Strengths and Weaknesses." *Sloan Management Review*, Spring 1976, p. 65.

[38]Adapted from Pearce II, John A., Robinson, Jr., Richard B. *Strategic Management*. Homewood, IL: Richard D. Irwin, 1982, p. 156.

Additionally, through the Baldrige Award, the government is encouraging better manufacturing.

Product lead time. Management of time, specifically lead time, has served as a catalyst for many companies.[39]

To gain control over lead time, it is necessary to identify the existing flow of material and information and to separate lead time into the following components:

- *Order Review and Release*—includes the preparation time of an order.

- *Queue Time*—the time during which inventory is at a work center waiting to be processed.

- *Run Time*—the time during which a machine is actually producing product

- *Move and Wait Time*—the time to move goods to the next work center. This depends on: proximity, plant layout, inspection procedures, or type of material.

Producibility. For a company to compete in DoD contract procurement, producibility measurement has to be part of the proposal process. The company may incur additional costs if producibility is not addressed prior to submitting a proposal.

Producibility measurement identifies:[40]

- problems that can arise during production

- subcontractors' abilities and deficiencies

- design requirements that achieve cost-effective production

Manufacturing processes. Manufacturing methods developed in recent years are changing the production process. The extent of their adoption will be the key factor for most companies to remain competitive. Factory automation has introduced the use of methods such as numerical-control machines, transfer machines, robots, automated warehouse systems, and material-handling systems.

By studying and understanding the capabilities and deficiencies of its present manufacturing processes, the company can gain important input into the manufacturing plan.

---

[39]Grauf, William M. "Lead Time Management, The Missing Link Between MRP II and JIT." *P&IM Review with APICS News,* August 1990.

[40]Department of the Navy—Best Manufacturing Practices Program, *Producibility Measurement for DoD Contracts,* 1990.

Tooling. The early 1980s saw a major push toward complex, highly automated manufacturing systems. These systems were predicated on the idea of putting as much state-of-the-art technology into a factory as possible. However, experience has shown that few installations can justify the cost of such systems. In order to assess the tooling and equipment capabilities within the factory, the company must understand such things as:

- current level of technology in the factory
- technology required to process materials of new product
- amount of equipment unique to processes already in place
- required level of unique tooling
- compatibility among tools
- maintenance requirements of tools and equipment

Facilities. Frequently, the importance of the facility which houses the manufacturing process is overlooked. This tends to occur because machines, tools, and a work force—and how to make them interact most efficiently—tend to be foremost in the minds of most production managers.

Facility capabilities play a major role in manufacturing planning. The following items help determine the facility's capabilities:

- life-cycle stage of the facility (e.g., initial planning and start-up, incremental expansion, maturation and reinvestment, renewal or shutdown)
- products manufactured and their output levels
- plant's capacity, manufacturing facilities, and technological capabilities
- specific process technology used and work flow
- production scheduling and control systems used
- interrelationships between this facility and other facilities, as well as with suppliers, the distribution system, and ultimate customers
- provisions for subsequent expansion and development of the facility

Human resources. Many manufacturing experts view human resources as one of the most critical components in manufacturing today and for the future. Analyzing direct and indirect staffing requirements in detail for the specific products is important. This analysis will help to identify alternative opportunities for improvement.

Staffing, the number of workers and their mix of their skills, and managing human resources are critical factors to counter increasing competition in many industries.

Training capabilities. Companies must regard their employees as their most significant asset and provide good general orientation as well as training in specific skills. The sixth rule in Deming's way *Out of the Crisis* is "institute training."[41] Nearly every conceivable improvement to manufacturing performance in some way depends on its acceptance by the work force. The proper training can greatly facilitate the implementation of change.

Training requires planning, change requires training and resources. Management needs training to learn about the company, from incoming material through to the customer. Also, management must understand and act on the problems that affect the production worker in carrying out the work with satisfaction.

Lessons learned. Past experiences can provide insight into what has worked before and what has not worked. The specific changes in business practices, methods, organization, and supplier relations must not be treated as individual items a list from which companies can pick and choose at will. The company can also use the lessons learned by other companies in their industry.

Companies must recognize their uniqueness before they copy someone else's success story. They must tailor the manner in which they incorporate technologies and processes to fit their own requirements— not someone else's.

Role in marketplace. The strategist must assess the company's role, such as "market leader" or "new entrant" in the marketplace. The strategist should compare the company's role to that of the competition.

Competition is at the core of the success or failure of many companies. Competition determines the appropriateness of a company's activities that can contribute to its performance, such as innovations, a cohesive culture, or good implementation. Understanding a firm's own capabilities helps to determine its ability to compete.

Competitive benchmarking is one tool for assessing a company's role in the marketplace. Competitive benchmarking can be done by comparing the company's profile and company products to those of their competitors. Some company-profile items for comparison are: market share, financial strength, diversity of competitors, technological strength, credibility, and

---

[41]Deming, W. Edward. *Out of the Crisis.* Cambridge, MA: MIT Center for Advance Study, 1982, p. 52.

manufacturing strategy. Some product issues for comparison are: product differences, pricing, performance, reliability, and quality.

Competitive benchmarking should precede the setting of manufacturing priorities. Knowing the value of what needs to be done and the pace required for accomplishment are essential elements.[42]

Strengths and weaknesses. A strength is any factor that provides the company a distinct competitive advantage. It is more than merely what the company has the competence to do. It is something the company does (or has the capacity to do) particularly well relative to existing or potential competitors. The importance of *strengths* rests with the unique capacity they give an organization in developing a comparative advantage in the marketplace.[43]

A weakness is any factor that the company does poorly, or does not have the capacity to do while the capacity does exist for key competitors.

Performance measures. Performance measures can provide a good picture about where the company is with respect to competition, use of available technology, and whether it is improving or not. A complete understanding of the entire manufacturing process can reveal a great deal about variation and quality in the process.

Process capability studies help to determine the inherent behavior of a process. The ultimate objective of a process capability study is to determine whether the existing process is capable of meeting objectives, quality and productivity goals, engineering targets, and customer needs. Typically, these studies will evaluate such factors as materials, machines, process, operators, time, controls, cost, management, environment, and schedule against their desired values.[44]

Constraints. For any company, resources such as facilities, skills, and money are limited. An explicit recognition of constraints or limitations is a prerequisite to a genuine understanding of the manufacturing problems and opportunities.

For example, production volume can be a major obstacle to the implementation of quality systems and automation. When only a few parts will be produced, little capital is made available for purchasing automated systems and up-to-date test equipment. In addition, other benefits resulting from learning curves and quality feedback may not be

[42]Sheridan, John H. "What Makes A Winner?" *Industry Week,* May 21, 1990, p. 34.

[43]Thompson, A. A., Strickland, A. J. *Strategy and Policy: Concepts and Cases.* Plano, TX: Business Publications, 1981, p. 54.

[44]Keyser, Jack. "Manufacturing Process Control," in *Microelectronic Reliability,* ed. Edward B. Hakim, Norwood, MA: Artech House, 1989, pp. 215, 219-221.

TABLE 7   Key Constraints

Economics of the Industry
Labor, burden, material, depreciation costs
Flexibility of production to meet changes in volume
Return on investment
Number and location of plants
Critical control variables
Critical functions (e.g., maintenance, production control, personnel)
Typical financial structures
Typical costs and cost relationships
Typical operating problems
Barriers to entry
Pricing practices
"Maturity" of industry products
Importance of economies of scale
Importance of integrated capacities of corporations
Importance of having a certain balance of different types of equipment
Ideal balances of equipment capacities
Nature and type of production control
Government influences and regulations

Technology of the Industry
Rate of technological change
Scale of processes
Span of processes
Degree of mechanization
Technological sophistication
Time requirements for making changes

possible. Long lead times, high levels of inventory, and poor levels of quality feedback usually result.

Table 7 lists a few constraints that the project manager must consider.[45]

Cost accounting. Many of the traditional cost accounting practices and systems have not been adapted to new business strategies. As a result, many costing systems are obsolete and can provide incorrect information. Incorrect or misleading costing can lead to erroneous assumptions and extremely expensive, misguided solutions.

---

[45]Skinner, William. "Manufacturing—Missing Link in Corporate Strategy." *Harvard Business Review,* May/June 1969, p. 144.

Traditional cost accounting systems ignore many product costs as general, administrative and self-insurance costs. We must change our perspective of how we view costs. We can no longer allow traditional financial accounting to influence our thinking as it has done in the past.[46]

For example, if a process is automated in an attempt to reduce the costs of direct labor, the overhead costs of such things as tooling, engineering, and maintenance increase. Traditional costing systems show managers a large investment and little or no return. This effect has caused confusion and dissatisfaction with investing in manufacturing technology. The reason for this is that cost accounting does not link the indirect costs to the product.[47] [48]

Make-or-buy. Make-or-buy analysis provides a technical review of a contractor's internal manufacturing capabilities and evaluates the subcontractor or vendor capabilities to provide certain products. One factor in the analysis is the impact of the in-plant loading on the overhead rates.

It is important to develop a make-or-buy plan that identifies major assemblies or components to be manufactured, developed, or assembled in-house and those which will be obtained from a subcontractor.

Make-or-buy programs are generally required when the work is complex, the dollar value is substantial, and price competition is lacking. On government contracts, for example, prospective contractor make-or-buy program information is required for all negotiated procurements except when the proposed prime contract:[49] [50]

- is estimated to be less than $2 million

- is for research and development, unless the contract is for prototypes or hardware and reasonable anticipation that there will be additional quantities of the product

---

[46]Sourwine, D. A. "Improved Product Costing: A Look Beyond Traditional Financial Accounting." *Industrial Engineering,* July 1990, p. 37.

[47]O'Guin, Michael C., "Activity-Based Costing: Unlocking Our Competitive Edge," *Manufacturing Systems,* December 1990, 8 (12), pp. 35–40.

[48]Keys, David E., "Limitations of Cost Accounting in an Automated Factory," *Computers in Mechanical Engineering,* July/August 1987, pp. 26-29.

[49]Acker, David D. and Young, Sammie G., LTC, USA, *Defense Manufacturing Management Guide for Program Managers,* 3rd Ed., April 1989, Ft. Belvoir, VA: Defense Systems Management College, pp. 10-4, 10-5

[50]Schwartz, Walter H., "Make It or Buy It?" *Assembly Engineering,* August 1989, pp. 27-29.

- involves only work that the contracting officer determines is not complex

Another exception is when the contracting officer determines that the program pricing is competitive and is based on established market prices or prices set by regulation.

Make-or-buy decisions for a specific program must be consistent with the overall strategic objectives.

Compare objectives

Successful improvement depends on determining those opportunities that will satisfy requirements and meet internal business objectives. To select opportunities wisely and fully realize the benefit of these requirements:

- Establish reasonable expectations based on needs and capability.

- Make a documented commitment stating the scope and time frame for improvement.

Identify and understand current initiatives. For some time, US manufacturers have been involved in some form of improvement to prepare for increasing globalization. Unfortunately, many firms simply tend to pick and choose initiatives to implement. Before deciding to embark on new initiatives, an organization should:

- identify the current initiatives in the company and their status

- have a clear understanding of what the initiatives are and are **not,** as well as what their intentions are

Coopers & Lybrand performs an annual study—the "Made in America" series—on the competitiveness of US manufacturers. The survey looks at the pace at which manufacturers are adopting new philosophies. In 1988 for example, about 14% of American manufacturers had moved to and implemented continuous improvement programs.[51] The survey also shows that by 1993, about half of US manufacturers will have moved toward JIT/TQM philosophies to support continuous improvement.

Investment in manufacturing technology is another initiative. Many executives feel that technology implementation (e.g., automation) will cure their manufacturing problems. When Deloitte & Touche surveyed

---

[51]Johnson, Henry J. "Preparing for Accounting Changes." *Management Accounting,* July 1990, p. 37.

759 senior executives of North American manufacturing firms, the executives admitted having little experience in advanced manufacturing technology. Yet, they expressed some interesting opinions:[52]

- 29% believed that use of the most advanced technologies yields *significant* benefits

- 48% credited advanced technologies with *moderate* or *minor* benefits

- 23% said advanced technologies yielded no benefits at all

These results confirm that understanding the initiatives at hand is critical to successfully implementing them.

Assess desired improvements

Once an organization has assessed current capabilities, compared its objectives, and understands those options currently available, then it can begin to identify the improvements required to compete.

> Many companies are responding to the competitive challenge by first restructuring their manufacturing operations, substituting a leaner, more efficient, and highly competitive organization.[53]

A common theme of restructuring is "back to basics," which focuses on identifying and implementing often dramatic changes leading to durable improvements in long-term profitability. Companies may falter in their attempt to restructure by resorting to "slash and burn" management and make across the board or arbitrary cuts without a clear understanding of exactly what cuts ought to be made, how deep they should be, and specifically what performance improvement should result in what period of time.

Well-managed companies, on the other hand, are conducting major reviews of their products and manufacturing processes and comparing these to their strategic objectives. They are examining which products are being made, using which processes, in which plants, for which markets. One of the goals of these reviews is to identify opportunities for standardizing production processes, equipment requirements, and end-item product characteristics. These opportunities will then translate into lower manufacturing costs.

According to an Ernst & Young study, successful companies showed these four characteristics:[54]

---

[52]Sheridan, John H. "The New Luddites?" *Industry Week,* February 19, 1990, p. 62.

[53]Deloitte & Touche. "Issues in Competitive Manufacturing." 1987, p. 1.

[54]Sheridan, John. H. "What Makes A Winner?" *Industry Week,* May 21, 1990, p. 30.

- Broadly focused planning—"Winning businesses were more likely to address matters of internal organization and external competition than counterparts."

- Broader product lines—Not only did the more successful firms have broader product offerings, but they worked harder at upgrading their products with frequent innovations. Also, they were more likely to be involved in international markets.

- Relevant performance measures—Awareness that traditional cost-accounting and performance-measurement approaches can be roadblocks to success, the better companies emphasized such measures as lead-time management, work-flow balance, and resource utilization.

- Improvement initiatives—In the more successful companies the accent is on quality improvement rather than cost reduction. Ironically, Mr. Ozan of Ernst & Young observes, "we found that the companies whose point of focus was quality actually achieved more cost reduction than those companies whose point of focus was cost reduction."

Use of cost management in decision making. Cost management combines elements of management accounting, production, and strategic planning. Tasks associated with cost management are cost accounting, product costing, measuring operational factors of production, and management accounting. An increasing emphasis on rapid development of low-cost, high-quality products in short intervals combined with manufacturing as a competitive advantage has changed the nature of the information required to manage the manufacturing process.

Cost management practices affect several areas in manufacturing and product management:[55]

- addressing manufacturing costs as part of the overall product life-cycle costs

- justifying the costs of advanced manufacturing systems and technologies

- understanding the links between business operations, manufacturing, and resource consumption

- understanding the relationship between the costs and savings realized from implementing quality improvement programs

- identifying performance measures that help shape current thinking and prepare for future improvements

---

[55]Cooper, Robin, "Introduction," in *Emerging Practices in Cost Management,* Ed. Barry Brinker, NY: Warren, Gorham, & Lamont, 1990, pp. xiii–xvii.

Activity-based costing. Activity-based costing (ABC) is a recent development in the area of cost accounting. Where conventional cost accounting focuses on allocating costs to each unit of a product, ABC focuses on the *activities,* or transactions, apply to a product during the manufacturing process. ABC also directs interest to cost drivers that are not directly related to unit-level characteristics. Some typical cost drivers are setup hours, number of setups, number of parts, and material-handling times. Studying these drivers help trace costs that tend to distort overhead costs in traditional systems.

ABC systems are more complex than traditional systems, and require a great deal of process analyses and design choices to work properly. However, when properly implemented, they provide new methods for costing products, modifying behavior, and focusing management attention on strategic issues.[56]

Because ABC systems focus on the activities in an organization or process, they also provide an opportunity for identifying areas of potential improvement and the measures to assess improvement. Figure 13 summarizes many of the differences between ABC systems and traditional systems.[57]

Identify programs for improvement. In the results of the 1987 Manufacturing Futures Survey, both European and US manufacturers ranked quality, high product performance, and delivery schedule performance as their top competitive priorities.[58]

Manufacturing improvement activities must be consistent with the manufacturing strategy. In addition to assessing present capabilities in terms of how they affect customer satisfaction and product costs, an organization must:

- identify process-simplification opportunities

- rank improvement opportunities based on customer satisfaction and business objectives

- identify quality-improvement projects to pursue

It is important to recognize that among the improvements identified, all improvements may not be doable within the time frame and budget.

---

[56]Cooper, Robin, "Elements of ABC," in *Emerging Practices in Cost Management,* Ed. Barry Brinker, NY: Warren, Gorham, & Lamont, 1990, pp. 3–23.

[57]McNair, C. J., "Interdependence and Control: Traditional vs. Activity-Based Responsibility Accounting," in *Emerging Practices in Cost Management,* Ed. Barry Brinker, NY: Warren, Gorham, & Lamont, 1990, pp. 421–430.

[58]Dumas, Roland A., Cushing, Nancy and Laughlin, Carol. "Making Quality Theories Workable." *Training & Development Journal,* February 1987, pp. 30-35.

	Activity-based accounting	Traditional accounting
Basic assumption	Interdependence	Independence
Focus	Organization	Individuals
Objective	Analysis	Cost Control
Control emphasis	Activities	Costs
Control point	Process	Outcomes
Variance usage	Improve process	Balance ledger/accountability
Standards	Historical/trended	Engineered/static
Goal encouraged	Continuous improvement	Meet standard
Control characteristics	n-Dimensional	One-dimensional
	Ambiguous	One-to-one map
	Strategic	Budget-based
	Financial and operational	Financial

Figure 13 ABC vs. traditional cost accounting.

Establish priorities. The following is a checklist of actions for assessing desired improvements:[59]

- Review and group the products and functions that affect manufacturing processes.
- Develop and evaluate factory-simulation models.
- Review manufacturing-support systems to identify which activities add value and which drive costs.
- Critically review the organizational issues.
- Balance management-performance measures.
- Nurture a *continuous improvement* philosophy.
- Keep desires consistent with objectives, goals, and *reality*.

External analysis. There is a definite need to find out what is new in technology, what the competition is doing, etc. Using the information from competitive benchmarking companies must identify how to use technology through partnerships and technology transfer.

The defense of America depends on high technology to detect and deter potential aggression. Sophisticated weapons require sophisticated production processes to assure reasonable cost and superior quality. Such manufacturing capability demands ongoing investment in new plants and equipment.[60]

---

[59]Sheridan, John F. "What Makes a Winner?", *Industry Week*, May 21, 1990, pp. 30, 34.

[60]Aerospace Industries Association. *Restoring Old Glory: A Strategy for Industrial Renaissance.* Washington, D.C.: Aerospace Industries Association of America, 1988, p. 1.

Industrial modernization incentives program (IMIP). Until recently, it was difficult for defense contractors to allot adequate capital for factory modernization. The DoD acknowledged the constraints and developed the industrial modernization incentives program (IMIP) to encourage contractors to accelerate plant-modernization plans.

The IMIP is a joint venture between the government and industry to accelerate the implementation of modern equipment and management techniques in the industrial base.[61] IMIP evolved from the Technology Modernization (TECH MOD) and Army Industrial Productivity Initiative (IPI) programs. IMIPs are implemented where competitive market forces are insufficient to motivate independent contractor modernization and where significant benefits such as cost reduction, elimination of production bottlenecks, improved quality, reliability, maintainability, and improved surge capability can be expected to accrue to the government.

IMIP objectives are to:

- reduce defense cost and lead times and to increase the quality of manufacturing through productivity gains
- produce a strong industrial base for surge and modernization requirements should a conflict or war arise

Manufacturing technology (MANTECH). MANTECH is a program for establishing, validating, and implementing advanced manufacturing capabilities to improve: producibility, productivity, cost reduction, and quality assurance.

MANTECH is an initiative to address potential areas of innovation and change. MANTECH focuses on advancing state-of-the-art manufacturing technologies and processes from the research and development environment to the production and shop-floor environment. Technologies with generic application required for defense systems and having high technical and financial risk characterize the candidate projects for MANTECH.

MANTECH projects demonstrate production application of emerging technologies. MANTECH is aimed at making first-class manufacturing process and equipment improvements in the production environment. Proven technologies resulting from the MANTECH program are candidates for implementation under IMIP. The MANTECH program can complement and support IMIP efforts. Figure 14 shows the IMIP and MANTECH relationship.[62]

---

[61]Westinghouse. *Westinghouse/AFSC partners in IMIP*. Westinghouse Brochure, p. 1.

[62]DoD 5000.44G. *Industrial Modernization Incentives Program*. Washington. D.C.: Assistant Secretary of Defense, 1986, pp. 1-7.

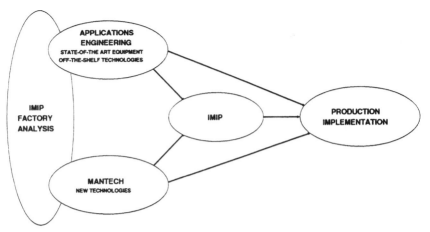

Figure 14 Relationship between IMIP and MANTECH.

Technology transfer. Developing and implementing technology within the company often requires investment of considerable time and resources. If technology is available, the company should use it. Technology transfer enables companies which are new to the market or without sufficient funding to develop the technology to enter the market and compete effectively. For example, companies funded under IMIP are required to make technologies they develop available to the DoD and its contractors.[63]

Another excellent example of technology transfer and information exchange can be found in the Navy's Best Manufacturing Practices (BMP) Program. BMP began in 1985 with the desire that DoD contractors identify and share their experiences (both good and bad) in order to improve quality in their industry. The BMP program sends survey teams to company sites to identify best practices, review problems, and document the results. The surveys are published in such a fashion that the issues are discussed and no proprietary or confidential material is released. Additionally, there is an annual workshop where industry participants share and discuss industry-wide issues. Another benefit of the workshop is the personal level of interaction between attendees. Outputs from the BMP program include a set of recommended solderability guidelines and a handbook on producibility measurement.[64]

---

[63]Westinghouse. *Westinghouse/AFSC partners in IMIP.* Westinghouse Brochure, p. 12.

[64]For more information on the BMP program, contact the BMP Program Director, Ernie Renner, at the Office of the Assistant Secretary of the Navy, Product Integrity (Research, Development & Acquisition).

Identify potential partners. Using one strategic objectives as a base, it is often wise to identify potential business partners before bidding on a program. Setting up strategic alliances allows a few firms to work towards their mutual benefit by allowing their respective strengths to work towards a common goal without spending resources to develop them. One member may contribute technical expertise, another can provide a strong production base, and yet another financial stability.

In some cases the company selects partners to get a competitive edge and in other cases the government may actually mandate it. For example, on the A-12 program, the Navy required stealth-capable firms to team with companies experienced with carrier-qualified aircraft.[65] The Advanced Tactical Fighter (ATF) program is a good example of two different teams developing prototypes in the competition to win a contract.

Develop improvement plans. Before undertaking any major improvement, companies must realistically analyze the opportunities, intentions, and goals. This requires analysts to apply their past experience, imagination, visualization capabilities, and creativity. Those who have worked in other plants, companies, and industries can cross-relate technologies and develop conceptual plans.

Bid decision

Now that the opportunity has been identified, and an assessment of capabilities and assessments has been completed, a critical question must be asked at this time:

> Does bidding on the project fall under the business unit and corporate strategies?

This question is very important. For a contractor to bid on a project—or for any company to go after a line of business, implies that the organization believes the opportunity will turn out to be a profitable venture over time. The question is intended to help identify the potential risks involved with the cost of pursuing and (hopefully) producing a product.

If the answer to this question is consistently "No," then it is imperative that the company re-evaluate its mission and ask such questions as:

- Why has it chosen this line of business?
- Does management really understand what it takes to compete and survive in this area?

---

[65]"Why the A-12 was Cancelled," *Jane's Defence Weekly,* February 9, 1991, p. 175.

- Can the organization successfully change its direction, with the present set of capabilities and resources, and be successful in another area?

Should there be a balance between pursuing and not pursuing an opportunity, then it is perfectly reasonable to continue to look for other opportunities, rather than try to force an issue. However, if a trend of "No's" begins to appear, then it may be prudent to ask the questions listed above.

Assuming the company has decided to pursue an opportunity and bid on a project, the development of its manufacturing plan begins to take on structure and become more formalized.

STEP 3—Develop Manufacturing Plan

Manufacturing planning activities should not occur in a vacuum. The manufacturing plan is a plan of action that must be consistent with both the manufacturing strategy and corporate strategy. A manufacturing plan ensures thorough planning for manufacture of the system to be produced. It documents the planning process sufficiently to enable review and traceability. Companies should apply the planning process iteratively throughout the life of the program.

> The manufacturing plan is the compilation of those documents a contractor uses to plan and manage their manufacturing effort.[66]

A manufacturing plan provides the means to link manufacturing strategies to business goals and to measure the progress in achieving these goals, through:[67]

- translation of business goals into quantifiable manufacturing objectives

- development of manufacturing concepts for operations, organization, equipment, product, facility, and information systems to meet manufacturing objectives

- definition of tactical programs necessary to implement manufacturing concepts

- construction of an implementation plan to manage tactical programs and measure progress in achieving business goals

---

[66]DI-MISC-80074, *Manufacturing Plan,* June 30, 1986, p. 1.

[67]Danzyger, Howard. "Strategic Manufacturing Plan? Your Competitiveness Depends on It." *Industrial Engineering.* February 1990, p. 19.

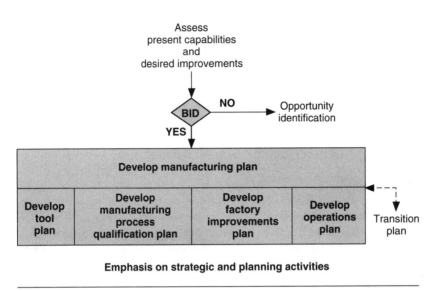

Emphasis on strategic and planning activities

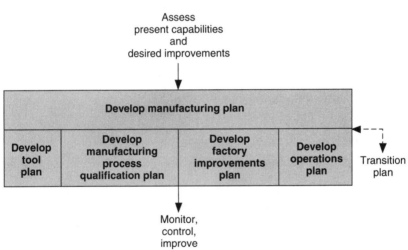

Emphasis on planning and implementation activities

Figure 15 Develop manufacturing plan.

When Does Manufacturing Planning Begin?

Manufacturing planning activities should follow a logical, iterative progression that starts as early as possible and results in the issuance of the manufacturing plan itself. However, it must be stressed that the manufacturing plan remains a living document.

Manufacturing planning begins in some form as soon as an opportunity is identified. Step 2 addressed many of the activities that begin at that time. Once the company decides to bid on a project, the development of a manufacturing plan should begin in earnest. Concepts such as concurrent engineering, total quality management (TQM), and producibility engineering and planning (PEP) all emphasize the early involvement of manufacturing in the development process. *Design to Reduce Technical Risk* addresses these concepts in greater detail. In addition, design reviews play a major role in assessing a product's production readiness.

Manufacturing plan in perspective

Manufacturing process planning and design is part of an overall concurrent engineering approach. It is important to understand how the manufacturing plan fits into the bigger picture. The manufacturing plan is closely linked with several other areas and plans for a given program. Some examples of program plans that are either influenced by, or influence the manufacturing plan are:

- statement of work (SOW) and contract data requirements list (CDRL)
- work breakdown structure (WBS)
- transition plan
- systems engineering management plan (SEMP)
- configuration management plan (CMP)
- risk management plan
- master test and evaluation plan (MTEP)

To make a program successful transition across all phases of development, there must be connectivity and feedback between all disciplines. Concurrent engineering philosophies dictate a multifunctional, integrated approach to development of products and processes. Plans are intended to guide activities and measure progress.

The configuration management plan describes the methodology and system which allows for the integration of efforts to proceed in an organized, reasonable fashion. The transition plan is the road map: the set of directions which provide structure to the program's life cycle.

It is critical that the planning efforts be dynamic. When any given plan is written once, regardless of the development phase, greatly increases the probability of failure. Plans need to be living; they must constantly be re-evaluated to reflect the current state of affairs and to address opportunities for improvement.

Manufacturing management

Manufacturing management has been defined as:

The techniques of planning, organizing, directing, coordinating, and controlling the use of people, money, materials, equipment, methods and processes, and facilities to manufacture systems.[68]

In the *Manufacturing Management Officer's Handbook,* the challenge facing Manufacturing Officers was to identify their responsibility to the project's Program Manager with the following objective in mind:

Try to minimize and clearly identify all the risk associated with the manufacture and production of a system.[69]

This objective can only succeed if the customer and contractor are able to work together as partners, and not adversaries. Additionally, identifying and minimizing risk should be a concern for all people involved with all phases of *any* project.

Often, in the case of such items as the WBS, there is an emphasis on discrete tasks and items rather than the process. In order to properly manage the manufacturing planning activities, a manufacturing management system needs to be in place.

Issues facing manufacturing management. A study conducted by the Manufacturing Management Council of the Society of Manufacturing Engineers concluded that there are six critical issues facing manufacturing management that managers must understand if they are to initiate change in their organizations. The issues identified by SME are:[70]

- Operations and Strategic Planning: This involves concentrating efforts and resources to find the right things to do. Included in these tasks are planning, management control, planning and analyzing investments, and cost estimating and control.

- Leadership Skills: Managers must develop and upgrade their leadership skills in order to properly understand the importance and communication of the company's philosophies and cultures.

- People: Managers must aid in the development and coordination of their most valuable resource—the work force. Other related concerns

---

[68]MIL-STD-1528A, *Manufacturing Management Program,* September 9, 1986, p. 3.

[69]Dwyer, James P., Major, USAF, *The Manufacturing Management Officer's Handbook,* Air Command and Staff College Report 85-0730, Maxwell AFB, AL: ACSC/EDCC, April 1985, p. 4.

[70]Veilleux, Raymond F. and Petro, Louis W., *Tool and Manufacturing Engineers Handbook, Volume 5: Manufacturing Management,* Dearborn, MI: SME, 1988, pp. v-vi.

in this area include labor relations and occupational safety and health concerns.

■ Manufacturing's Function: The manufacturing function needs to be a management partner with other functions such as engineering, finance, and marketing. Manufacturing needs to be involved with design and standards efforts. Additionally, manufacturing managers need to understand the roles of manufacturing, in relation to such things as JIT, CIM, and project management.

■ Facilities: Manufacturing managers need to understand the importance of using existing facilities more efficiently. Responsibilities in this area include facilities planning, equipment planning, production and inventory control, and materials management.

■ Quality Management and Planning: Managers must understand that quality cannot be mandated, and that it is an outcome of outstanding management, systems, employees, and dedication. Also, manufacturing quality efforts are part of the overall company-wide quality policies, tools, and improvement efforts.

Scope of manufacturing management. Manufacturing management is a big job. It requires integration, communication, and cooperation from all levels of a program, from the systems program office (SPO), down through to the subcontractors. Strong interaction with the customer facilitates the understanding of manufacturing implications on cost and schedule.

Manufacturing management strongly emphasizes early involvement and continuous improvement. Early on, its main objective is to identify and reduce manufacturing risk. Manufacturing risk is the risk that the system will not be manufactured to requirements within the cost, schedule, and performance constraints of the program.

In determining manufacturing risk, such items as producibility, manufacturing processes, tooling to be developed, testing and special test equipment, and logistics must be considered.

The manufacturing management program continually assesses production readiness. Production readiness is the degree to which a program is ready to proceed into production. A program is ready for production when a producible design is complete and the managerial and physical preparations necessary for initiating and sustaining a practical manufacturing effort allow a production commitment to be made without causing unacceptable risks of impact to schedule, performance, cost, quality, reliability, maintainability, or other established thresholds.

Production readiness review (PRR) is a formal examination of a program to determine if the design is ready for manufacturing, if manufacturing engineering problems have been resolved, and if the contractor has adequately planned for the production phase. The review may be conducted incrementally. *Design to Reduce Technical Risk* discusses PRRs and associated reviews in greater detail.

Manufacturing management example. MIL-STD-1528A, *Manufacturing Management Program,* provides an excellent example of the type of structure that is required of a good manufacturing management program. Table 8[71] summarizes many of MIL-STD-1528A's requirements.

Configuration management and change control

Since the manufacturing plan is part of a hierarchy of documents that span the life cycle of a product, it is especially critical that it be maintained and continuously updated. Manufacturing can be a major milestone in the acquisition process, but it should never be an afterthought.

Inputs to manufacturing come from design, test, risk analyses, the SOW and WBS, and logistics. Without a proper mechanism in place to ensure that design changes are properly documented, manufacturing risk can never be mitigated. Similarly, changes due to manufacturing decisions must be communicated if the other organizations are to successfully deliver a producible design. *Design to Reduce Technical Risk* addresses the topics of design release and configuration control in greater detail.

Subcontractor relationship

An input that is often overlooked in manufacturing planning is the subcontractor. Subcontractors and vendors should be part of the development team and are an integral part of the manufacturing process. A good working relationship with vendors can lead to inventory reduction, reduced work-in-process, reduced scrap and rework, increased floor space, reduced costs, and an overall increase in quality.

Teamwork and communication with vendors regarding strategy, planning, policy, and implementation issues can also facilitate the make-or-buy analyses. Part 2, Parts Selection and Defect Control, and Part 3, Subcontractor Control, discuss many of the strategies and activities related to both vendor and part selection.

[71]Acker, David D. and Young, Sammie G., LTC, USA, *Defense Manufacturing Management Guide for Program Managers,* 3rd Ed., April 1989, Ft. Belvoir, VA: Defense Systems Management College, p. 10-2.

TABLE 8   MIL-STD-1528A Manufacturing Management Requirements

Management Area	Requirement
Planning	Identify and Obtain Production Resources
	Identify and Resolve Risk
	Identify and Obtain Capital Commitments
	Identify and Obtain Tooling and Test Equipment
	Verify Manufacturing System
	Integrate Program and Factory Planning
	Integrate Make-or-Buy Analysis
	Integrate Industrial Material Management
Design Analysis	Producibility Analysis
	Process and Methods Analysis
	Design and Manufacturing Engineering Integration
	Production State-of-the-Art Analysis
Operations Management	Production Scheduling and Control
	Work Measurement
	Manufacturing Surveillance
	Control of Subcontractors and Vendors
System Manufacturing Assessments	Manufacturing Feasibility
	Manufacturing Capability
Contractor/Government Interface	Manufacturing Management Program Review
	Manufacturing Management/Production Capability

A sub-contractor quality improvement program. General Dynamics Corporation Electronics Piece Part Integrity-Product Improvement Program is a part of an overall corporate quality initiative intended to improve the overall quality of GD's products and increase the productivity of their processes.

The policy is intended to improve piece part quality *before* fabrication and assembly, thus building in quality. A common objective across all divisions of the corporation is major improvement in throughput and end-item quality and reliability. The program has five main features:

- Establishing an electronic piece parts quality improvement program (QIP) with selected suppliers. The primary objective of the GD/supplier QIP is increased supplier product quality.

- Performing component evaluation and reliability stress testing for defectives. The objectives of this activity are to assure high quality components for manufacturing and provide early feedback to suppliers. As quality levels are achieved and maintained, the levels of testing may be reduced or eliminated.

- Implementing a closed loop feedback program. Timely feedback assures rapid corrective action by suppliers and GD. Feedback also provides data to either confirm a supplier's commitment to their quality goals or to identify areas for corrective action.

- Establishing a supplier performance evaluation and control program. The main objectives of the source control program are to assure constant quality and reduce testing such that random sampling can be done with little risk.

- Establishing a subcontractor flowdown plan. Also referred to as the "Black Box Flowdown Plan," this program promotes the involvement of subcontractors and the customers in a QIP to improve the quality levels of "black boxes" that are assembled into GD's systems.

Understand manufacturing issues

Attempts to plan what could be referred to as the manufacturing process raise several issues and subprocesses that are important to identify and define. In addition to producibility, other factors that must be considered in the manufacturing process include:[72]

- Part Fabrication and Assembly Process: Part fabrication refers to the actual forming of parts into their desired shapes. The assembly process details the sequence of steps required to put a product together.

- Tolerances: Design tolerances affect each step in the manufacturing process. Tightness of dimensioning and tolerancing requirements greatly influences assembly sequence and product cost.

- Feeding and Handling of Parts: Manufacturers need to consider the handling, transportation, and orientation of parts within the assembly process. If a part is easy to grasp and orient for assembly, it is more likely to be placed accurately.

- Testing and Inspection: Testing and inspection can help manufacturers to deliver high quality parts, as well as provide data for statistical analyses. Complexities in assembly and fabrication increase the likelihood of faults and defects. Testing of parts, subassemblies, and final assemblies can identify commonly occurring faults and opportunities for improved quality control and process improvement.

---

[72]Nevins, James L. and Whitney, Daniel E., *Concurrent Design of Products and Processes,* NY: McGraw-Hill, 1989, pp. 231-277.

These issues, and many more, must all be considered and addressed during the development of the manufacturing plan.

Contents of a manufacturing plan

A manufacturing plan is a compilation of a number of documents that detail the system and factors necessary to achieve an effective, efficient manufacturing system.

At different stages of a product life cycle, the manufacturing plan emphasizes different aspects of the manufacturing management task. Early on, the plan attempts to describe a method to produce, test, and deliver a system within time and cost constraints. The plan tries to detail how the method would work, given the present set of facilities, tooling, and personnel constraints. As the final Production Readiness Review (PRR) approaches, the manufacturing plan becomes more detailed, describing the entire set of manufacturing operations. Issues addressed in the plan include the manufacturing organization, make-or-buy plans, resources and manufacturing capability, and detailed production plans.

The actual contents of a manufacturing plan can vary greatly. The Data Item Description (DID) for manufacturing plans (DI-MISC-80074), lists 27 different types of information that may appear in a manufacturing plan. The Application chapter of this part discusses the manufacturing plan for the F-16 Fighting Falcon, produced by General Dynamics.

Sub-plans.   The manufacturing plan also addresses a hierarchy of sub-plans it addresses. Four of the major sub-plans that are tied to the manufacturing plan are:

- Tool Plan
- Manufacturing Process Qualification Plan
- Factory Improvements Plan
- Operations Plan

Develop tool plan

There is often a great deal of confusion as to what the terms tooling, special tooling, and special test equipment (STE) really mean. Many programs refer to needing "tools and equipment" without a full understanding of what is actually required.

> Tool planning encompasses those activities associated with a detailed, comprehensive plan for the design, development, implementation, and

prove-in of program tooling based upon a visible corporate policy and structured around a documented practice.[73]

Tooling. Tools can be thought of as the equipment (e.g., machines, machine tools, instruments) required for producing a system. Examples of tools are drills, lathes, milling machines, and robots. Generally, tools are considered those items that can be acquired off-the-shelf and require little or no modification for producing the system.

Special tooling is defined as:

> All jigs, dies, fixtures, molds, patterns, taps, gauges, other equipment and manufacturing aids, and replacements thereof, which are of specialized nature that, without substantial modification or alteration, their use is limited to the development or production of particular services.[74]

Basically, special tooling is the set of tools that must be developed for a specific program or system. Special tools, by nature, are generally more expensive than regular tools and require significant planning and design efforts. An example of special tooling may be a custom end-effector for a robotic assembly application. In this reference guide, "tooling" will refer to special tooling.

Types of tooling. Tools are generally thought of as either fixed or flexible (alternately, "hard" or "soft"). Each type has an area of specialization and economic impact that must be considered as early in the process as possible.

Fixed tooling is generally designed for a particular task, or a task where there are few changes. Generally, fixed tooling is used for high-volume production or preplanned flexibility. Flexible tooling is reconfigurable for a wide variety of tasks, and can be controlled and adapted by an operator or sensors.[75]

Special test equipment. *Special test equipment,* also referred to as factory test equipment, are those testing units that are essential to the testing and inspection of a particular product during production, from unit assemblies up to the final product. STE items do not include special tooling and general plant testing items. STE can include custom-designed test fixtures and inspection equipment.

---

[73]NAVSO P-6071, p. 6-36.

[74]DSMC, p. B-17.

[75]Nevins, James L. and Whitney, Daniel E.,*Concurrent Design of Products and Processes,* NY: McGraw-Hill, 1989, pp. 285-289.

Types of special test equipment. STE has two main functions: product inspection and test, and process verification. As with tooling, these groups have areas of specialization and economic impact that must be considered as early in the process as possible. Product inspection and test examines an item in manufacturing to evaluate it visually and functionally for defects. STE can test for such things as tolerances, manufacturing defects, fit and finish, and function to ensure that the manufacturing process has not adversely affected the product. Process verification examines how the process itself is performing. Measurement of such things as tool wear, throughput, and yield help the company to determine potential areas for improvement.

Testing of materials in process requires early consideration of testability in all aspects of a product's design. *Design to Reduce Technical Risk* addresses the issues of design for test and built-in test in greater detail. STE also provides valuable data on product and process during manufacture. Step 4 and Part 2, Parts Selection and Defect Control, discuss the importance of process monitoring and control as well as reinforcing the notions of continuous improvement.

Strategic considerations. Production tooling and test equipment may require a significant investment in time, money, and resources. Production of specialized equipment is a development effort that requires the same type of up-front consideration and planning as that of the entire system. Design, verification, and production of tooling and test equipment must be concurrent and closely linked to the system development effort.

Custom tools and equipment can cost a great deal of money. Production rate, material selection, number of different tools required, and design complexity are the main cost drivers. Critical decision aids are trade studies, make-or-buy analyses, producibility assessments, and risk assessments. These help to identify problems early, when they are less likely to increase costs significantly.

Figure 16 illustrates the tool/equipment selection process.[76]

Role of tooling and STE. Traditionally, tooling was considered to be only the machines and machine tools that were required for such jobs as metal cutting and forming. With technologies ranging from composites to micro-electronics to computer vision, the need for customized tooling and test equipment has increased.

---

[76]Veilleux, Raymond F. and Petro, Louis W., *Tool and Manufacturing Engineers Handbook, Volume 5: Manufacturing Management,* Dearborn, MI: SME, 1988, p. 19-7.

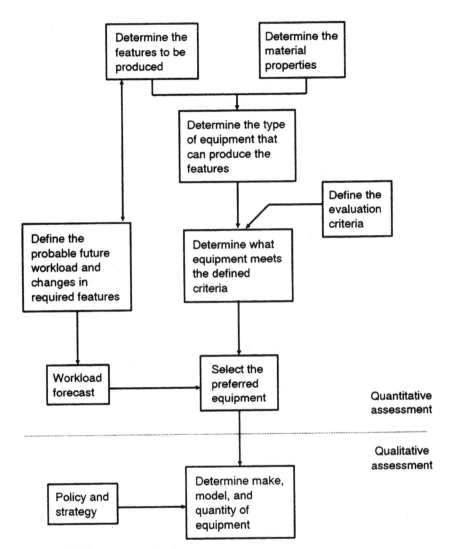

Figure 16 Tool/equipment selection process.

By definition, tooling and STE are unique to a given process or product. Concurrent engineering philosophy calls for an integrated design of product and process. Tool planning and design *must* be part of any concurrent engineering effort.

Products and processes are driving the need for early definition of the requirements of specialized equipment. Other parts of this guide address how issues such as process planning, process qualification, and factory improvements relate to manufacturing planning. Tool planning must be an equivalent and integral function of the manufacturing effort as well.

Tooling and STE provide the means for creating and assuring the product design and the process design.

A tooling lesson learned. An example of the lesson learned from not being able to design or understand the need for special tooling and test equipment can be found in the A-12 Avenger program's cancellation. Defense Secretary Dick Cheney said that the prime contractors were unable to "design, develop, fabricate, assemble, and test the A-12 aircraft within the contract schedule."[77]

Problems with dealing with composites, a major item in stealth designs, was the main reason the A-12 was 18 months behind schedule and $2.7 billion over budget. The prime contractors had limited experience with building large composite structures, and literally had to develop the technology as the program went along. Additionally, the Navy's principle deputy general counsel in reporting on the A-12, concluded that "the contractors underestimated the level of effort needed to stabilize the aircraft design, as well as the impact this would have on tooling and fabrication. In addition, they overestimated their internal capacity for tool design and fabrication of metal-composite components."[78]

The Application chapter describes the tool planning strategy and activities employed by Bell Helicopter for many of their designs.

Develop manufacturing process qualification plan

Methods such as design for assembly (DFA), design for manufacturability (DFM), producibility engineering and planning (PEP), Taguchi methods, robust design, and concurrent engineering stress the early integration of product and process design. Yet, a highly manufacturable design is effectively worthless if the manufacturing process to produce it is not available.

A manufacturing line is *qualified* if it can be characterized completely in terms of:

- capability
- staffing
- sources and amount of variation

---

[77]NYT, January 6, 1991, p. 1.

[78]Braham, James, "What Price Stealth?", *Machine Design*, February 21, 1991, 63(4), p. 27.

- conformance with standards (e.g., ISO 9000, ANSI, DoD directives, MIL-STD's, tri-service regulations)

- production rates

Qualifying the manufacturing process improves manufacturing performance by ensuring:

- built-in quality

- consistent and predictable quality and reliability

- provisions for real-time analysis and continuous improvement

- faster changes

By successfully focusing on a process involving both design and manufacturing engineers, a company is ensured of producing products in volume, on time, and within cost. Design engineers should consult with manufacturing engineers to further refine the process using production equipment. The successful blending of engineering and manufacturing can be facilitated by expert systems, engineering work stations, and integrated databases.

Process qualification.   Production sample is one means of qualifying the manufacturing process. It enables a company to obtain information about the manufacturability of an item, using actual production equipment and processes. The feedback from building the production sample can be used to improve the design and manufacturing processes. It also enables a manufacturable product at a lower cost and higher quality.

Training is an important ingredient in implementing the qualified manufacturing process. Design and manufacturing engineers determine the training needed in the production environment. Designers should follow the product out to the production floor and perform many of the training functions.

Important steps in process qualification are: process characterization and process capability studies.[79]

Process characterization.   Process characterization involves:

- identifying the steps in the process

- listing variables present in each step

- listing the normal values for variables that have control over the process

---

[79]Keyser, Jack. "Manufacturing Process Control," in *Microelectronic Reliability,* ed. Edward B. Hakim, Norwood, MA: Artech House, 1989, pp. 215, 219-221.

A process can be characterized using process flowcharts, cause and effect diagrams, and determining current process parameter settings. Flowcharts document the steps in the process, identify the inputs to the process and outputs of the process. Cause and effect diagrams determine the parameters acting on each step of the flowchart and factors that may or may not affect the output.

Process capability studies. Process capability concerns the process variation and the variables affecting quality and productivity of the process. Process capability studies determine:

- whether the process can meet objectives, such as design specifications, customer needs, etc.

- the inherent or natural variation (behavior) of a process, identify and eliminate causes underlying any unnatural behavior

The factors internal to the manufacturing process are: materials, processing machinery and test equipment, the process itself, time-related, operators, and automatic controls embedded in the process.

The external sources of variation are those external to the manufacturing process itself. They may come from within the company or from an external customer. They are powerful driving forces and may have a significant impact on the product quality. Some of the most significant external forces are cost, management, schedules, and environment.

In addition, process qualification enables the following:

- Process optimization—conducting studies and experiments to determine preferred levels of process parameters

- Process and product control and monitoring—establishing the new level of performance, obtaining control, regular monitoring to maintain control

- Process improvement—adopting continuous improvement as a practice

A process qualification program. Upon recognition of the fact that its competitive environment was characterized by powerful competitors, sophisticated designs, rapidly changing technology, and increasing customer expectations, the Westinghouse Electronic Systems Group (ESG) examined its focus on the product/process interface.

ESG created Producibility Assurance Centers (PACs) and Transition Assurance Centers (TACs). The PAC/TAC idea is to foster the integration of the design and manufacturing processes. The basic theme of the PACs is to emphasize flexibility, creativity, and innovation in product and process design. The TACs serve to represent the initial pilot production lines to validate the manufacturing processes prior to starting full-rate production.

In ESG's Advanced Interconnect Technology Laboratory, the PAC/TAC strategy achieved a 50% reduction in the time to move to next generation of printed wiring packaging technology from development to production.[80]

Sample manufacturing process qualification checklist. Table 9 presents a general checklist used to qualify manufacturing processes according to a set of certification requirements. The checklist is often tailored to specific processes as needed.[81]

Qualified manufacturers list program. To reduce the cost of electronic components, semiconductor manufacturers and the government are participating in a joint effort to qualify the manufacturing processes for monolithic microcircuits (under MIL-I-38535) and for hybrids (under MIL-H-38534/MIL-STD-1772).

The qualified manufacturers list (QML) effort takes advantage of the near-zero defect levels many semiconductor suppliers achieve. The QML program recognizes that incoming inspections are costly and are not needed when suppliers implement SPC programs to ensure quality is built in. The objective of QML is a 10-fold to a 100-fold decrease in the price of silicon microcircuits.[82]

With QML, the manufacturing processes are certified rather than individual parts, as in current Qualified Parts List (QPL) and MIL-M-38510, General Specification for Microcircuits. The traditional approach of certifying parts under the Joint Army–Navy (JAN) programs and under the MIL-M-38510 program was costly, lengthy, and inefficient.[83] Parts became obsolete almost as soon as they were qualified. The lengthy audit process had to be repeated with each upgrade. QML eliminates the need to re-qualify parts from a certified line.

The key element of QML is in-process monitoring of the manufacturing processes to ensure device yield and reliability.

---

[80]Teixeira, John, "Concurrent Engineering," paper presented at the 4th Annual Best Manufacturing Practices Workshop, September 11, 1990, Scottsdale, AZ.

[81]Personal communications with C. J. Keyser, AT&T Bell Laboratories.

[82]Burgess, Lisa. "Thomas: Pushing the Pentagon Toward QML." *Military and Aerospace Electronics,* February 1990, pp. 39-40.

[83]Gardner, Fred. "Hold Down Ballooning Costs and Boost Quality." *Electronic Purchasing,* June 1988, p.57.

TABLE 9   Manufacturing Process Qualification Checklist

MANUFACTURING PROCESS QUALIFICATION

Certification Requirements

	A	M	I
Manufacturing flowchart with identified inspection checkpoints, input and output variables (parameters)			
Experienced/trained operators, certified where necessary or desirable			
Experienced/trained maintenance personnel			
Appropriate supervision			
Knowledgeable about process and product requirements			
Certified inspectors			
Adequacy of facilities			
Facilities (equipment) maintenance program			
Calibration system			
Assembly instructions, work instructions, or standards			
Inspection procedures			
Latest issues of drawings and specifications available			
Incoming material and parts—inspection and control			
Traceability, serialization, or lot identification			
Control of nonconforming product			
Inspection results recorded, maintained, and used for process improvement			
General housekeeping			
Overall compliance with quality plan			

A = Adequate   M = Marginal   I = Inadequate

The benefits of the QML program are:

- better control of manufacturing process
- better use of facilities
- fewer government audits and lower qualification costs
- predictable part costs
- improved delivery schedules

A product should not be released to the production floor until there is sufficient proof that the product is manufacturable. The desired end result of a qualified process is predictability by repeatability. In general, developing a manufacturing process qualification plan is a good business practice because it ensures cost-effective programs and operations.

The Application chapter discusses the case study of AT&T Micro-electronics' manufacturing facility in Allentown, PA, which was the first facility to have its manufacturing processes qualified.

Develop factory improvements plan

Factory improvement does not just mean updating the equipment in the factory. It involves striving to achieve an integrated and support-ive mix of layout, material flow, inventory control, manufacturing pro-cesses, maintenance, and plant control. An organization must be able to assess the value-added of the technology before investing capital.

> ...a productive factory is not a collection of isolated machines or auto-mated systems; it is a single, integrated effort.[84]

Factory of the future. The factory of the future has traditionally been viewed as a high-tech, automated "lights out" operation with few or no human operators. This misconception has led to many companies invest-ing a great deal of capital in order to automate processes they do not understand. As far back as the 1950s, organizations sought to put in new systems based on new, unproven technology without fully understand-ing the scope of their efforts.[85]

High-tech does not necessarily mean replacing human operators with robots. It means re-thinking processes and determining the best mix of people, tools, machines, and resources. It means combining this understanding with product design in order to determine the most effi-cient and cost-effective manufacturing processes.[86]

Computer-integrated manufacturing. The concept of CIM is covered in greater detail in *Design to Reduce Technical Risk*. The basic idea in devel-oping a CIM architecture is that a quick fix will not necessarily provide lasting improvements. CIM requires an organization to address several key issues:[87]

---

[84]Fife, William J. Jr., "The Automation Imperative," *Assembly,* 1990 Buyer's Guide Issue, July 1990, p. 216.

[85]Havatny, Josef, "Dreams, Nightmares, and Reality," *Computers in Industry,* 4(2), 1983, pp. 109-114.

[86]Royce, WIlliam S., *Is Manufacturing Obsolete?,* Business Intelligence Program, Report No. 83-800, Menlo Park, CA: SRI International, 1983, pp. 10-15.

[87]Harmon, Roy L. and Peterson, Larry D., *Reinventing the Factory,* NY: Free Press, 1990, pp. 233-236.

- Product Engineering: It is critical to understand and define the links between design engineering and its tools (e.g., CAD, CAE, GT).

- Process Engineering: To properly generate plans and instructions, manufacturing engineers need to work out the interactions between CAM, CAPP, and factory equipment.

- Manufacturing Planning and Control: Planning and execution of master production schedules require a proper understanding of such concepts as MRP II and JIT.

- Factory Support: Resources must be available to support scheduling, tooling, and maintenance.

- Factory Execution: Coordination and control of machines, operators, and cells facilitate the production of quality products.

- Automation: Identify areas where automation is needed, such as high-volume repetitive functions and material storage and retrieval systems.

- Information: Connectivity and integration require clear channels of communication between people, machines, and management.

Automating a problem still leaves a problem. Companies must understand the processes, simplify them, automate where needed, and then integrate the solution.

Maintenance. Decisions regarding manufacturing equipment must be made with consideration of the equipment's reliability. Manufacturing can not be competitive if the production line continually has problems. There is a strong need to rediscover preventive maintenance.

Traditionally, maintenance was focused on preventive and corrective practices. While practical and valuable, preventive maintenance is not enough when things like time to market and JIT are considerations. Manufacturers need to account and plan for predictive and diagnostic maintenance also. Planning for the maintenance tasks is discussed in the Develop Operations Plan section.

Maintenance responsibility does not just belong to the manufacturing organization. Suppliers also should take an active role in designing and proving their equipment prior to installation in a factory. Accordingly, manufacturers should include the suppliers as part of their factory improvement planning team.

Material handling, flow, and inventory. Initiatives and techniques such as that JIT, FMS, cellular manufacturing, automated material storage/retrieval systems, CAPP, and GT strive to provide solutions which address concerns of material inventory, supply, handling, and flow.

Automation of the production can have many payoffs; automation of the waste before and after the production makes little sense.[88]

The key point to remember about material handling is that it must efficiently *support* the manufacturing process. Material must be able to move through the manufacturing facility so that the entire environment works smoothly. Material handling can help to reduce inventory, control inventory, and free space on the factory floor. If material is not handled well process efficiency suffers. Damaged parts, for example, increase the amounts of scrap and rework.[89]

Plant layout. The physical layout of a factory directly influences the efficiency of the production system. In a JIT system, for example, the facility is designed to minimize work-in-process, material handling, and cycle time, while improving feedback. The manufacturing process, equipment considerations, and material handling concerns must all be included in the layout planning function.[90]

Whether undertaking a re-layout of an existing facility or looking for opportunities to improve the present facility, the company has to understand how the layout currently functions. To get a firm grasp on how the layout affects processes, physical and economical constraints as well as practical limitations have to be considered. Areas of analysis include product volume and mix, production processes, production departments and tasks, material flow, and space requirements. Additionally, layout decisions must offer benefits to both the company and the employees, not just the manufacturing bottom line.[91]

Plant operations. Building control has been included in the past 20 years in the integration considerations of manufacturing facilities. Heating, ventilation, air-conditioning, waste management, and safety systems often have a direct influence on the production processes. For example, in electronics manufacturing, temperature, humidity, and airborne particles have direct impact on the process yields. Integrating

---

[88]Shonberger, Richard J., *World Class Manufacturing*, NY: Free Press, 1986, p. 45.

[89]Bergstrom, Robin P., "Some Things Just Don't Translate Well," *Production*, 102(12), December 1990, pp. 54-57.

[90]Lubben, Richard T., *Just-In-Time Manufacturing*, NY: McGraw-Hill, 1988, pp. 119-121.

[91]Usher, John S. and others, "Redesigning An Existing Layout Presents a Major Challenge—And Produces Dramatic Results," *Industrial Engineering*, June 1990, 22(6), pp. 45-49.

these plant operations with manufacturing operations can help reduce wastes, control downtime, and diagnose problems.[92]

JIT: A misunderstood philosophy. The Japanese success in implementing the just-in-time (JIT) manufacturing control philosophy has led many American firms to view it as a "must do" to be successful in manufacturing. Many companies have implemented JIT without fully understanding what it really means. JIT's basic philosophy is simplification through waste elimination. Waste is anything that does not add value to the process or product. JIT attempts to identify all non-value-added activities and remove them. JIT is *not*:[93]

- an inventory program
- an effort that involves suppliers only
- a cultural phenomenon
- a materials project
- a program that displaces MRP
- a panacea for poor management

JIT and MRPII. Manufacturing resource planning (MRPII) and its predecessor, materials requirements planning (MRP), deal with material scheduling, inventory management, capacity planning, and shop floor control (both are discussed in relation to CAM and CIM in *Design to Reduce Technical Risk*. There is a misconception that JIT and MRPII systems cannot work together. In fact, a properly designed MRPII system supports JIT by facilitating planning and identifying requirements prior to implementation.[94][95]

For example, it is not "JIT" when JIT parts delivery is achieved by having a vendor set up a warehouse in the parking lot.

Figure 17 illustrates the activities in a comprehensive JIT manufacturing system.[96]

---

[92]Benassi, Frank, "Honeywell Integrates Building and Process Controls in Factories," *Managing Automation,* March 1991, 6(3), pp. 38-40.

[93]Veilleux, Raymond F. and Petro, Louis W. *Tool and Manufacturing Engineers Handbook, Volume 5: Manufacturing Management,* Dearborn, MI: SME, 1988, pp. 2-18.

[94]Goodrich, Thomas, "JIT & MRP CAN Work Together," *Automation,* April 1989, 36(4), pp. 46-48.

[95]*The Determinant Factors for a Successful MRPII Implementation,* Saratoga Springs, NY: Business Education Associates, 1987.

[96]Veilleux, Raymond F. and Petro, Louis W. *Tool and Manufacturing Engineers Handbook, Volume 5: Manufacturing Management,* Dearborn, MI: SME, 1988, pp. 2-19.

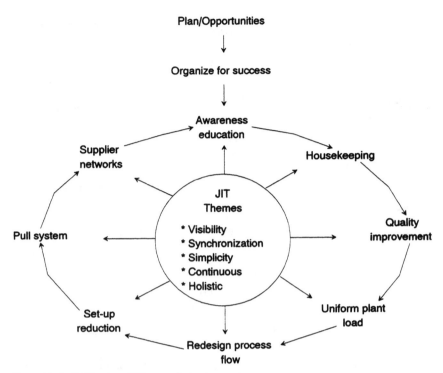

Figure 17 Activities in a JIT manufacturing system.

Develop operations plan

The operations plan helps companies to manage shop floor processes. The plan helps fill the gap between industrial engineering and production. Often referred to as shop floor control or plant floor management, the operations plan connects the actual execution of functions with the data required for successful manufacture and measurement.

> In a totally predictable production environment, demand forecasts would always be realized; bills of materials would be absolutely accurate; suppliers would ship their orders on time and with total accuracy; nothing would be misplaced or miscounted in the storeroom; machines on the shop floor would never fail; all manufacturing personnel would be present when expected; and all intervals in the process would be fully predictable.[97]

Process planning. A process plan is the detailed description of how raw material is transformed into a finished part. Process planning is the

---

[97]Doshi, Bharat T. and Krupka, Dan C., "Integration of Planning and Execution Operations: Theory and Concepts," *AT&T Technical Journal,* July/August 1990, 69(4), p. 91.

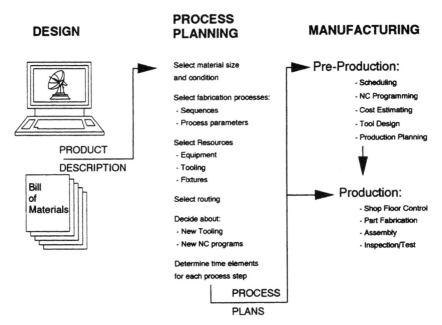

Figure 18 Process planning interfaces.

method used to describe a part (e.g., dimensions, tolerances, materials), identify the manufacturing steps, and map the product design with the production process.[98] Figure 18[99] illustrates the function of process planning. *Design to Reduce Technical Risk* defines, discusses, and gives examples of process planning and computer-aided process planning (CAPP).

The process plans provide the basic instructions for manufacturing a product. The plans are very detailed, exactly specifying such things as how to set up and operate a machine, work instructions, time standards, bills of materials, engineering drawings, inspection, part classifications, parameters (e.g., feed and speed of cut), cost estimates, and reference documents.[100] [101]

---

[98]Bhaskaran, Kumar, "Process Plan Selection," *International Journal of Production Research*, 1990, 28(8), pp. 1527-1539.

[99]Adapted from Sutton, George P., *Computer-Assisted Process Planning*, Business Intelligence Program, Report No. 765, Menlo Park, CA: SRI International, 1988, p. 2.

[100]Gould, Lawrence., "Putting a CAPP on CIM," *Managing Automation*, August 1990, 5(8), pp. 17-19.

[101]Metz, Sandy, "Making Manufacturing Better, Not Just Faster," *Managing Automation*, August 1990, 5(8), pp. 22-24.

The process plans provide cost and time estimates to the shop floor. Once the process is up and running, these estimates can be compared to the actual values, thus evaluating the estimates and processes and identifying potential areas for improvement.

Work instructions. The DoD refers work instructions through MIL-HDBK-50A, which stated "all work affecting quality shall be described in clear and documented instructions of a type appropriate to the circumstances."[102] Documenting instructions for areas such as assembly, fabrication, processing, inspection, and test helps ensure that production occurs in the fashion prescribed by the process planning. As with requirements, it is critical that the instructions be clear, timely, and concise. Additionally, they must provide qualitative and quantitative criteria when appropriate.

Operators generally perform tasks from sets of step-by-step instructions, illustrations, and references to other documents. Often, each instruction is independently generated (either manually or by computer) and maintained so that it is difficult, if not impossible, to retrieve a complete document at any given time. Because of this, it is vital that the information be managed effectively throughout the process.

To assure quality, work instructions must be well coordinated and communicated throughout the facility. It is critical that the instructions on the floor match the design of the product currently in production. The wrong set of instructions during manufacturing could severely affect product quality, reliability, and overall cost. A complete configuration management system allows coordination not only during development, but also in production.

Paper or paperless? A fully integrated work instruction system should include:[103]

- documentation that is readily available at the operators station
- illustrations integrated with instructions
- control and accountability
- synchronized distribution of documentation
- support for traceability and audit of the manufacturing instructions

---

[102]MIL-HDBK-50A, *Evaluation of a Contractor's Quality Program,* June 26, 1990, p. 11.

[103]*Report of Survey Conducted at Lockheed Missile Systems Division, Sunnyvale California,* Best Manufacturing Practices Program, OASN-PI(RD&A), August 1989, p. 10.

Traditional systems are paper-driven. There is a push towards paperless operations using central repositories for data and operator work stations. The Application chapter presents a case study where Northrop converted the F/A-18 assembly line to a paperless operation.

Maintenance. Maintenance is the ability to plan, manage, and control while maximizing the uptime of a facility's production processes. Maintenance provides manufacturers with the means of keeping their facilities in top-notch condition.

Successful and profitable companies include maintenance as a critical component of an effective manufacturing program. Operations plans must also account for maintenance of plant equipment. Good maintenance involves more than just fixing broken equipment. It also involves diagnosis and monitoring before a problem occurs. The "Don't fix it until it breaks" attitude has proven to be very costly.[104]

The benefits from a proper maintenance program include increased uptime and output; reduced scrap and rework; reductions in unexpected production costs; improved product quality and consistency; control of product and production costs; the ability to meet JIT manufacturing commitments and delivery schedule; reduced unexpected capital expenditures for equipment; and the extended life of capital equipment.

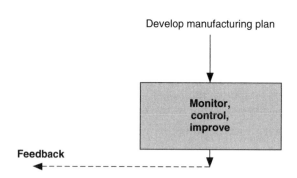

Emphasis on planning and implementation activities

Figure 19 Monitor, control, improve.

---

[104]Benassi, Frank. "Maintenance Management: Manufacturing's Final Frontier." *Managing Automation.* March 1991, p. 34.

Step 4—Monitor, Control, Improve

Good enough is never good enough. If improvement is not continuous, companies can fall behind.

Step 4 assumes that the program has transitioned into production. Various manufacturing processes are in place and need to be managed using tools and techniques currently available. Initiation of this step follows the implementation of the manufacturing plan.

Process-oriented management

Process-oriented management consists of monitoring, controlling, and improving existing processes to reduce variations and, when needed, trying new processes.

Process monitoring is a systematic procedure to detect, quantify, and eliminate the assignable causes of variability. Process control is the routine monitoring of a process against its statistical limits to detect out-of-control conditions and remove their sources. Process improvement is a structured activity to introduce fundamental changes that will achieve new levels of performance.[105]

Only by continuously reducing the variability of processes can companies remain competitive in today's marketplace. A significant management focus and frequent reviews are required to make continuous process improvement a reality.

Two aspects of continuous improvement that must be addressed to achieve long-term success are innovation and a relentless daily pursuit of reduced variability and improved levels of performance. Changes in process are most often created by those who are closest to the action and empowered to make a difference. It is the daily incremental improvements to existing processes on a continuous basis that are the foundation of long-term success.[106]

Process-quality management and improvement methodology

Process-quality management is planning and doing the activities necessary to sustain process performance and identify opportunities for

---

[105]Donnell, Agustus and Dellinger, Margaret. *Analyzing Business Process Data: The Looking Glass.* Indianapolis, IN: AT&T Customer Information Center, 1990, p. 96.

[106]Bryce, G. Rex. "Quality Management Theories and Their Application." *Quality,* February 1991, p. 24.

improving quality and reducing costs. Process-quality improvement is acting on opportunities to drive the process to a new level of performance.

There are many methods available to continuously monitor, control, and improve processes, such as the Process Quality Management and Improvement (PQMI) methodology developed by AT&T.[107]

PQMI is a methodology for building an effective process management system with *quality* as its foundation. The methodology supports an overall quality architecture. Within the architecture quality-related activities are linked across the boundaries of major functional organizations, such as marketing, sales, design, manufacturing, and distribution.

Principles of PQMI. The principles given below offer a structured approach for defining process goals and understand the *best way to achieve* them:

- Process-quality improvement focuses on the end-to-end process.

- Commitment to quality is one of prevention and continuous improvement.

- Everyone manages a process at some level.

- User needs drive process-quality improvement.

- Corrective action focuses on removing the root cause of the problem rather than on treating its symptoms.

- Process simplification reduces opportunities for errors and reworking.

- Process-quality improvement results from a disciplined and structured application of the quality-management principles.

Table 10 shows tasks to be performed at each of four distinct stages of the PQMI methodology.[108]

Ownership

For each process, there should be an owner who assumes responsibility. The owner should ensure that the planning is consistent with the business and corporate strategies. The responsibilities of the owner are:

---

[107] AT&T Quality Steering Committee. *Process Quality Management & Improvement Guidelines.* Indianapolis, IN: AT&T Customer Information Center, January 1989, pp. 71-126.

[108] AT&T Quality Steering Committee. *Process Quality Management & Improvement Guidelines.* NJ: AT&T Bell Laboratories, 1988, p. 13.

TABLE 10   Stages of PQMI

Stage	Purpose
Ownership	To ensure that someone is in charge of the process and that a team exists to carry out day-to-day quality-management activities
Assessment	To ensure that the process is clearly defined
Opportunity Selection	To understand how internal-process problems affect user satisfaction and cost
	To identify and rank order opportunities for process improvement
Improvement	To achieve and sustain a new level of process performance by implementing an action plan for realizing opportunities identified in the previous stage

- to manage the process across functional or organizational boundaries
- to establish a team that assumes ownership of major subprocesses of the cross-functional process
- to define roles and responsibilities of the team and ensure that people involved have ownership of the process performance

Assessment

Identifying process requirements enables the company to establish specific and measurable manufacturing needs. These requirements, together with program objectives, drive the process requirements. This may involve redefining the overall manufacturing cycle into a group of essential functions. Each function should represent the collective efforts of one or more departments to accomplish a major objective of the company.

For example, ensuring a steady, reliable stream of raw materials to manufacturing may involve coordination among the sales, purchasing, traffic, receiving, and accounting departments. Similarly, production reporting may involve interactions among manufacturing, quality, engineering, accounting, and MIS organizations.[109]

Understand the process. Managing a process requires a description in terms of:

---

[109]Kaeli, James K. "A Company-wide Perspective to Identify, Evaluate, and Rank the Potential for CIM." *Industrial Engineering,* vol. 22 (7), July 1990, p. 24.

- how the process actually operates

- the interactions among the customers, suppliers, and major work groups

- the process boundaries and interfaces

Flowcharts are useful in capturing the process flow visually in terms of functions, systems involved, and interfaces. For example, flowcharts depict the process, such as the movement of materials (raw, product, and nonproduct) as they are received and processed through the facility. The flowcharts should include general operational steps (including inspections), moves, delays, and storage. Process conditions should be recognized and recorded.

Monitor the process. Monitoring the process involves reviewing the documented flow of data and materials through each functional area. Representatives from each functional area should meet to evaluate this information collectively. Each functional area must be evaluated for existing strengths and weaknesses.

Most competitive operations today need accurate, timely data.[110]

Analyzing the process requires measures that accurately reflect what is expected of the process. In general, there are three measures for a process:[111]

- Quality—How well does the process meet the need?

- Timeliness—Was the process working when it was needed? Was it on schedule?

- Cost—Is the process worth what it costs?

Data collection and analysis. The steps in collecting and analyzing data are:

- define specific reasons for collecting data

- decide on measurement criteria

- ensure accuracy of the measurement system, including operator performance

---

[110]Eighth Annual Computer Survey. "Data Collection: The Key to Efficiency." *Modern Materials Handling,* November 1990, p. 66.

[111]Donnell, Agustus and Dellinger, Margaret. *Analyzing Business Process Data: The Looking Glass.* Indianapolis, IN: AT&T Customer Information Center, 1990, p. 14.

- ensure accuracy of measuring equipment
- analyze using several tools

Many sources of data come from the processes discussed in Step 3, during the development of the manufacturing plan. It is especially important to recognize the data types identified under tooling, process qualification factory improvements, and operations planning. Data to be gathered typically include such things as: machine downtime, work flow (including bottlenecks), operator times, amount of scrap and rework, inventory levels, material flow volumes, supplier delays, and cost. To obtain the data:

- The team develops the information-requirements list. The team submits the list to appropriate departments and operational groups.

- The team surveys existing operations to obtain information that is not available from existing sources.

- Team members interview key operating personnel, focusing on management perspectives and user-operational needs.

Analysis tools. There are many process-management tools, such as; cause-and-effect diagrams, histograms, Pareto diagrams, and others.

Cause-and-effect diagrams. A cause-and-effect or "fishbone" diagram represents relationships between a given effect and its potential causes. It is drawn to sort out and relate the interactions among the factors affecting a process and to determine the root cause of a problem.

Histograms. A histogram is a graphical representation of the frequency distribution of variable data. It is useful for visually communicating information about a process and for helping the company to make decisions about where to focus improvement efforts.

Pareto diagrams. A Pareto diagram is a simple graphical technique for rank ordering causes from most to least significant. It is based on the Pareto principle, which states that just a few of the causes often account for the most effect.

Control the process. Process data are of little value if they cannot be understood. When properly analyzed and presented, the data can be used to determine process capability, the degree of process control, and areas for improvement.

Many different methods, such as *statistical process control* (SPC) can be used to monitor and control the process. Statistical process control, also known as statistical quality control (SQC), is the application of statistical techniques to measure, analyze, and control the variability of a process.[112]

"Statistical thinking" can provide a powerful method of process monitoring and control. However, it is important to note that the analyses are only as good as the data, understanding, and interpretation of the SPC results. Seven fatal mistakes when using SPC are:

- *Weak management commitment*—management commitment is essential for the success of quality improvement.

- *Poor training*—all levels of management should receive adequate training to enable them to make decisions.

- *Poor measuring system*—without good measuring systems the data obtained will be erroneous.

- *Poor quality management system*—quality management system is basic for SPC.

- *Poor corrective action system*—without a good corrective action system it is difficult to manage the system when it goes out of control.

- *Remote control thinking*—this refers to use of SPC in an office, to control a process in the shop.

- *Poor success awareness*—everyone affected must be in on the plan.

With measures in place to track the key elements of manufacturing, gathering useful data helps assess how good the process is.

SPC is addressed in more detail in Part 2, Parts Selection and Defect Control.

Opportunity selection

Process management tools serve to identify the fact that there is a problem in a specific area of the process, but do not isolate the problem. Investigating the process to identify and understand the problem areas is essential to continuous improvement. Systematic investigation of processes often reveals significant opportunities for improvement.

Once problems have been identified, opportunities for improvement can be selected. Choosing the right area for improvement is extremely important. The following activities must be performed by the team:[113]

---

[112]Crosby, David C. "How to Succeed in SPC." *Quality in Manufacturing.* March/April 1991, 2(2), pp. 36-37.

[113]AT&T Quality Steering Committee. *Process Quality Management & Improvement Guidelines.* Indianapolis, IN: AT&T Customer Information Center, January 1989, p. 52.

- Review improvement opportunities.
- Establish the priority of each improvement opportunity.
- Negotiate objectives.
- Decide on important projects.

The team must also gain consensus on an opportunity before the improvement can begin to be implemented.

Improvement

Once consensus has been obtained, the following activities must be performed to improve the process:[114]

- Organize a process improvement team and develop an action plan.
- Determine root causes of the problem.
- Test and implement the solution.
- Hold process gains and follow through.
- Perform periodic process review.

After implementing selected improvements, return to the beginning of the methodology to identify new opportunities for improvement.

PQMI, or any process-oriented management methodology, is not a one-shot deal. It serves to emphasize and give structure to an environment of continuous improvement.

Figure 20 illustrates the basic structure of the continuous improvement strategy in place at Texas Instruments Missile Systems' HARM factory.[115]

---

[114]AT&T Quality Steering Committee. *Process Quality Management & Improvement Guidelines.* Indianapolis, IN: AT&T Customer Information Center, January 1989, pp. 57-58.

[115]Schuch, Linda K., "TI Does It Right the First Time," *Assembly Engineering,* September 1989, p. 13.

TOTAL QUALITY EXCELLENCE

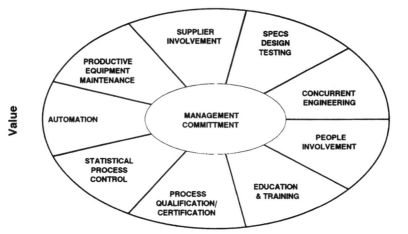

Producibility

Figure 20 Continuous improvement strategy at HARM factory.

Chapter

3

Application

This chapter presents the following case studies and examples to illustrate some of the principles discussed in the Procedures chapter:

- *"Greenfield Approach" to Manufacturing Strategy*: Details the strategic, planning, and implementation efforts of TRW in developing a new division.

- *Understanding Strategic Manufacturing Decision Areas*: Details eight categories and presents some typical questions to help evaluate each area.

- *Sample Manufacturing Strategy*: Presents the manufacturing strategy and planning statements from Unisys EISG.

- *F-16 Manufacturing Plan*: Summarizes the structure and contents of the F-16 Block 50 Manufacturing Task Analysis.

- *Tool Planning at Bell Helicopter*: Describes the tool planning strategy employed at Bell Helicopter.

- *AT&T's MOS V Qualified Manufacturing Line*: Describes the rationale and activities followed by AT&T in becoming the first company to make the DoD's Qualified Manufacturer's List.

- *Paperless Assembly Line for the F/A-18 Hornet*: Details the IMPCA system employed at Northrop Aircraft Division's Hawthorne, CA facility.

- *Air Force Lessons-Learned Program*: Presents examples from the abstracts found in the Lessons Learned database.

"Greenfield" Approach to Manufacturing Strategy

In the early 1980s, TRW identified and began to pursue the programs that would meet the high-tech avionics needs of the 1990s. Programs such as the ATF, LH, A-12, F-15, and F-16 require highly complex, integrated avionics that promoted the state-of-the-art. TRW desired a pre-eminent position in this field and recognized the need for processes that would allow cost-effective production of advanced, integrated avionics.

TRW decided to meet the challenge by developing an entirely new division and facility, the Military Electronics and Avionics Division (MEAD). MEAD was a strategic initiative based on TRW's strengths in modular, integrated avionics. MEAD is intended to meet avionics needs from concept development all the way through to the depot.[116]

A greenfield approach

TRW realized that the technical challenges facing avionics in the 1990s would require an innovative approach. The traditional approach of design, test, build, redesign, retest, rebuild was too slow, too unreliable, and too expensive. MEAD's development was based on the idea of a *greenfield approach.* The greenfield approach refers to the development of a facility from the ground up, where none has existed before.

To become a pre-eminent supplier of high-tech avionics systems, TRW identified several key differentiators to success:

- quality

- controlling the process, not the product

- cost

- design to cost

- schedule

- JIT manufacturing

- manufacturing technology

- capacity

A critical ingredient of the MEAD development strategy was TQM, with the cornerstone being concurrent engineering. This allowed TRW to simultaneously consider and integrate the needs of design, manufacturing, logistics, and the customer as early as possible. Additionally, it emphasized, "do it right the first time" and managing the process, not the statistics.

---

[116]This case is based on materials from several sources: J. Calkin's presentation at the 1990 BMP Workshop, BMP Survey conducted at TRW MEAD in March 1990, presentation materials from TRW (video, promotional literature, view graphs), and personal communications with P. Glaser and J. Calkins of TRW MEAD, Rancho Carmel, CA.

Background

From 1981 through 1984, TRW's activities were primarily limited to pursuing development contracts for integrated avionics. The conceptual design of a modern avionics manufacturing facility did not begin until 1985.

In 1985 and 1986, two studies were conducted by external consultants: one focused on the facility concept and the other on site selection. The concept study concluded the need for an advanced avionics manufacturing facility (AAMF). The study concluded that the factory should be focused, integrated, JIT, and possess state-of-the-art manufacturing capabilities. The site selection study examined many locations and performed assessments between co-location and such factors as labor and operational costs. The study concluded that by integration of marketing, engineering, and production would reduce start-up costs and risks. The resulting AAMF plans called for an integrated, world-class factory to be located in Rancho Carmel, California.

Corporate and top management granted approval for the go-ahead was granted in 1986, with implementation beginning in 1987. The initial schedule called for the plant to be operational in late 1989, and to have the first systems off the line in 1990.

Key strategic elements. Key strategic elements for developing the long-range plan included:

- an engineering-manufacturing interface to provide for concurrent engineering, design to cost, and producibility in design efforts

- the creation of design and manufacturing standards to assure consistent product design and to help eliminate cultural barriers

- the development of flexible, automated, fine-pitch surface-mount technology

- the establishment of an Advanced Manufacturing Technology Center (AMTC) to validate processes prior to factory floor implementation

- the inclusion of CIM in to the factory architecture

- the definition of a test strategy to accommodate the anticipated complexity of the hardware to be tested

In 1987, the implementation plan was modified due to the changing requirements of the DoD advanced-platform programs. A three-phased approach was implemented, emphasizing a facility that minimized economic and technical risks while focusing on the ability to support planned and surge capability requirements successfully. The first phase included the AMTC, which was already under way. The second phase called for the establishment of a lower-capacity initial production facility (IPF). This tactic reduced initial investments and provided greater flexibility to meet the ever-changing requirements of the DoD. The final phase entails building a full production facility once product and capacity requirements are better defined. The IPF would also provide capacity overlap once the new facility had begun. These changes pushed the schedule for initial production to late 1990.

In 1988, the strategy elements were further refined to reflect an implementation plan that would use existing facilities longer. A key part of the strategy required that the IPF achieve all the objectives of the full facility, but on a smaller scale. This would allow MEAD to evaluate options for full-scale production that would minimize investment and provide lowest cost products. Certification of the IPF was also to be pushed up to early 1990 to demonstrate manufacturing capabilities at the earliest possible date.

IPF focus.   Overall, the IPF focuses on designing process control into the manufacturing system. Other key aspects of the work conducted at MEAD include:

- an in-line SMT standard electronic module (SEM) assembly line
- understanding the process and design-related variables
- process control through statistical analysis
- mastering fine-pitch technology with a zero-defect philosophy

Process capability studies.   MEAD extensively used Taguchi methods to conduct its process capability studies.

Figure 21 illustrates TRW's process capability study methodology. The methodology reduces the amount of rework and inspection in a given process. Figure 22 part a[117] illustrates the conventional process flow in a traditional factory. In this case, 100% yield is a target, not a realistic goal. Conversely Figure 22, part b, illustrates the process flow in the factory the future. Here, the process is monitored and controlled, preventing the manufacture of defective products. Among the AMTC's goals are process-controlled lines and 100% yield.

Key facility elements

The AMTC was completed in 1987 and process development activities were completed in late 1988. Equipment was moved into the IPF from

---

[117]Wright, Dennis J. "Avionics Manufacturing in the 1990s," *QUEST,* TRW Space & Defense Sector, 13(1), Summer 1990, p. 6.

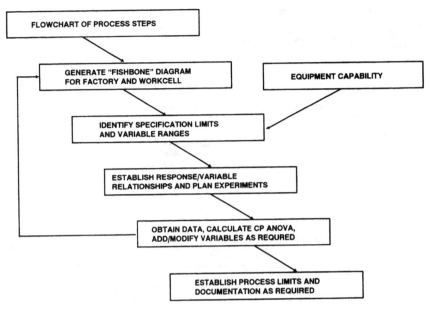

Figure 21   Process capability study methodology.

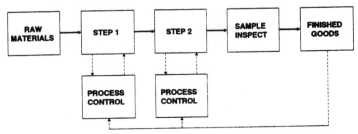

Figure 22   Conventional vs. process-controlled lines.

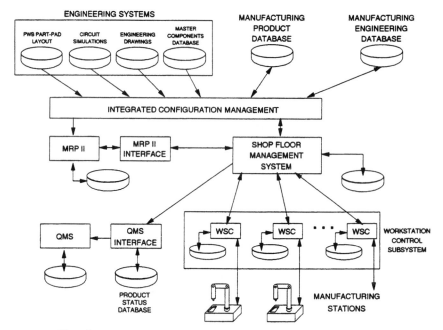

Figure 23 Shop floor control system.

the AMTC in 1989, and the facility became operational in 1990. One of the key features of the facility is a shop floor control system (SFCS) intended to coordinate business and engineering systems on the factory floor. The structure of the SFCS is illustrated in Figure 23.

A modular, automatic avionics test system (MAATS) was developed to provide design-factory-depot testing. MAATS allows for a family of configurable avionics test systems which are tri-services compliant.

Robotic assembly is used extensively to reduce handling and to increase the accuracy of the placement of fine-pitched devices.

Design. MEAD extensively uses CAD and CAE. Simulations are used to verify the functionality of the design. Layouts are used to generate parts lists and preliminary production plans. These lists and plans are then scrubbed by the design and manufacturing standards databases before baselining designs.

Configuration management. A configuration management system helps to maintain accountability and control of design and manufacturing data. Also, the system allows for the accurate communication of changes.

CAD/CAM. Designs are passed from CAD/CAE into the CAM systems to finalize and formalize manufacturing plans and assembly sequences. The plans are then checked for consistency with an optimal allocation of resources. The CAM system also provides one of the major links to the MRP II system.

Shop floor control. Design to cost and statistical analyses help qualify processes and define process parameters prior to production. Emphasis is placed on solder joint integrity, reliability, advanced technology designs, and surface-mount technology.

The shop floor control system helps to meet these goals on two levels: the work cell and the work station. The work cell level manages the data flow between the shop floor, MRP II system, and CAD/CAM. The work station level takes care of individual manufacturing operations. Each station on the floor has an icon-based touch-screen terminal which allows for the real-time monitoring and control of processes. A typical station extensively uses automation and computer-aided visual inspection.

Parts and materials. Parts and materials must meet specifications before entering the assembly sequence. A bar code tracking and control system helps critical information travel with parts and assemblies through the process. Typically, this information includes work instructions, quality requirements, and processing information (e.g., thermal profile for proper reflow soldering). This tracking system also allows defect data to be recorded for reporting to the customer or for process improvements.

Facility status

As of late 1990, the initial facility was producing FSD hardware for two programs. An outgrowth of MEAD's in-house process capability studies was a contract awarded for electronic manufacturing process improvement. This work with Wright-Patterson AFB focuses on developing a fully process-controlled manufacturing line. Efforts are in place to continuously improve processes such as advanced manufacturing technology, shop floor control, and MRP II.

To become to a full production facility depends on the requirements of MEAD's customers. Regardless of pending work, the IPF is intended to provide, at minimum risk and investment, the ability to win and perform on new contracts.

Understanding Strategic Manufacturing Decision Areas

Table 5 summarized the eight categories which Wheelright identified as critical for strategic manufacturing decisions. Briefly, the eight categories are:

- Capacity
- Facilities
- Technology
- Vertical Integration
- Work force
- Quality
- Production Control
- Organization

An explanation of each and some typical questions to help evaluate each area follows.

Capacity

In the face of growing global competition, the development of a capacity and facilities strategy is essential to any manufacturing enterprise.[118] This strategy comprises several elements including how much capacity to add, when to add it, and where to add it. Following are some questions regarding plant capacity:

- How much capacity should be available?
- How should the capacity be utilized?
- What is the correct mix of permanent vs. temporary capacity?

Facilities

To formulate a facility strategy, companies must understand market trends and forces well enough to plan plant capacity.
The following questions must be asked about facilities:

- Where should facilities be located?
- What are the number and sizes of the plants?

---

[118]Butler, Michael P. "Facility and Capacity Planning Using Sales Forecasting by Today's Industrial Engineer." *Industrial Engineering.* June 1990, p. 52.

- How should the factories be focused?
- Are the facilities adequate to the task?

Within any manufacturing facility, managers are continually forced to make trade-offs among competing objectives. Because the focused factory is responsible for only one specific product, the number of competing objectives are fewer, and trade-offs are made more consistently.

Technology

The "factory of the future," i.e., robots and lasers, CAD, CAM, CIM, and flexible machining centers all represent tremendous opportunities. However, American industry has made relatively little progress in terms of either large investments or, more importantly, large successes. The tools are there, but the progress is extremely slow.

In developing a manufacturing strategy, strategists must ask the following questions:

- What is the appropriate level of technology the company should employ?
- How much of the process technology should the company develop internally?
- How should the company maintain the equipment?

Vertical integration

Vertical integration is the ability to convert raw materials into finished products within one firm. A good example of a vertically integrated facility is a steel mill where items such as ore and coal enter and finished steel products leave.

Strategists must ask the following questions regarding the degree of vertical integration:

- Should the company make or buy required materials?
- How many suppliers should the company have?
- How should the company manage and control the supplier network?

Vertical integration addresses issues such as *make-or-buy*. Make-or-buy analysis is a technical review of internal manufacturing capabilities and evaluates the subcontractor or vendor capabilities to provide certain products. One factor in the analysis is the impact of the in-plant loading on overhead rates. This is especially important in the case of a facility that is involved with many programs, because the overhead rate of other programs can affect the overhead to be applied to the program of interest.

Make-or-buy decisions may be required if the work is complex, the dollar value is substantial, and price competition is lacking. Make-or-buy decisions of the strategy must follow the overall business strategy.

The make-or-buy plan identifies major assemblies or components that are to be manufactured, developed, or assembled in its facilities and those that will be obtained from outside sources. Contract manufacturing is one of the alternatives to doing it in-house.[119] Contract manufacturing capabilities range from printed circuit board shop to full product engineering and support. One reason for using such services is capital investment. The contract manufacturer spreads the cost of its facility, equipment, and services across its entire customer base.

Work force

Almost every company is trying new methods for managing its work force. Changes in industry are driving towards empowered teams, increased work force participation, and better management of employees.

The following are some questions addressing the needs of effective human resources management:

- What skill requirements are required?
- How should the company train its employees?
- What types of employee incentive and recognition programs are there?

Quality

Worldwide expectations and demands for quality have increased dramatically. Quality improvement requires change and a fundamental assessment of the way the firm does business.

Answering the following questions helps the company define a quality strategy:

- What is the company's quality policy?
- What approach should be taken to achieve quality objectives?
- What is the cost of quality?

---

[119]Schwarz, Walter H. "Make It or Buy It?" *Assembly Engineering.* August 1989, p. 27.

Production control

In the present manufacturing environment, there are pressures to reduce time to market, reduce inventory, and increase quality. At the same time, product offerings are expanding and the production environment is being recast due to technology and new operating principles. Regardless of these changes, the efforts of all production-related departments must be coordinated.

Unfortunately, managers tend to latch on to programs like JIT and MRPII, convinced by consultants, academicians, and text books that these methods will cure all ills. Too often, managers have bought two or three of those programs, along with some quality control, and the programs may pull the company in three different directions.

The following questions address some production control issues:

- How should the material needs be planned?
- Can the present set of vendors meet future needs?
- What system of production control should be employed?

Organization

Companies are making organizational changes to improve development cycles and inventory turnover. They also are starting to overhaul their antiquated accounting systems.

Strategists must address the following organizational questions while developing a strategy:

- How should the company be organized?
  Should it be organized around product or process?
- How many layers in management are required?

Also, many changes are occurring in vendor relationships. Managers are working closely with vendors—often with fewer of them. Automobile manufacturers are doing a tremendous job with this, despite many problems. They are taking vendors into their confidence and making them part of the family.

Sample Manufacturing Strategy

The Unisys Engineering Information Systems Group (EISG), as part of their overall Operations Manufacturing Plan for 1990, documented their mission statement and strategy as presented below:[120]

---

[120] Courtesy of Unisys Engineering Information Systems Group, St. Paul, MN.

Mission statement

The mission of EISG Operations is to be a low-cost, quality producer through our commitment to continuous improvement.

Manufacturing strategy

- Establish Centers of Excellence to focus technology, eliminate redundancy, and minimize cost.
- Improve resource utilization by transitioning volume production to low-cost sites and consolidating facilities.
- Promote synergy through process/procedure standardization and common systems.
- Strengthen business relationships with key suppliers to minimize cost, improve quality, and ensure timely material availability.
- Support EISG financial and management objectives through aggressive asset management and program performance.
- Establish a competitive culture through employee and team development, total quality management, and a commitment to continuous improvement.

F-16 Manufacturing Plan

General Dynamics' Ft. Worth Division is the birthplace of the F-16 Fighting Falcon. Over the years, the F-16 has gone through several revisions. These revisions necessitated modifications to the assembly line in order to accommodate the newer systems.

For the F-16 Block 50 Program, which began deliveries in May 1991, the manufacturing plan was written to be a living document. It is intended to be the basis for future revisions of the plan, the plane, or both.

Table 11 depicts the high-level structure of the Block 50 Manufacturing Task Analysis.

What follows is a summary description of the plan's contents.[121]

TABLE 11   Structure of the Block 50 Manufacturing Task Analysis

Block 50 Manufacturing Task Analysis
1.   General Dynamics Organization
2.   Resources and Manufacturing Capability
2.1   Existing Facilities
2.2   Facilities Arrangement
2.3   F-16 Production Layouts
2.4   Manpower Forecasts
3.   Manufacturing Plan

---

[121]Courtesy of General Dynamics Ft. Worth Division, Ft. Worth, TX.

Block 50 Manufacturing Task Analysis

1. General dynamics organization

This section describes General Dynamics' (GD) organizational structure for the F-16 program. This section presents organization charts for the F-16 Program Office, the production departments, and the Block 50 Production/Quality Assurance organization. The Block 50 Production Work Breakdown Structure is also presented here.

2. Resources and manufacturing capability

This section details the resources and manufacturing capabilities available for use on the F-16 program at the Ft. Worth location.

2.1 Existing facilities

This section characterizes Air Force Plant No. 4 in terms of its integrated capability for research, development, production, test, and support of weapons systems. Other areas covered here reflect the continuous maintenance and upgrading of the facility, personnel, and location.

2.2 Facilities arrangement

Figure 24 summarizes the production flow of an F-16. The plan also presents other figures that illustrate the integrated capabilities of the facility. The entire facility is

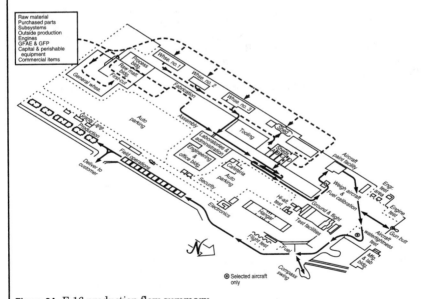

Figure 24 F-16 production flow summary.

(Continued)

also described in terms of such characteristics as number of buildings (127), total acreage (602), floor space (6.9 million sq. ft.) and ownership (e.g., GD or government).

2.3   F-16 production layouts

The document presents figures illustrating the layout of Air Force Plant No. 4 for the production program. The assembly areas are mapped out and layout is allocated according to production rates.

2.4   Manpower forecasts

Figures present the projections for manpower requirements and acquisition in the areas of tool engineering, tool manufacturing, and factory. There is room to project the needs of any new training requirements of personnel. However, since this modification required no new additional skills or processes, the plan called for previously proven training and certification of personnel.

3.   Manufacturing plan

This section of the Manufacturing Task Analysis describes the tooling and manufacturing tasks for the F-16 Block 50, and is based on the engineering statement of work. Concurrent production of a previous version of the F-16 will continue after the start of Block 50, necessitating the retention of tooling and manufacturing capabilities for the separate model. However, no major assembly line changes are required.

   Co-production of certain parts of the airframe are part of the overall program plan, and are detailed in a separate plan, the Make-or-Buy plan.

   The actual manufacturing, or production, plan is scheduled according to the Schedule Work Breakdown Structure. This part of the plan depicts the actual manufacturing sequence of the airframe's structure and the installation of the subsystems and avionics. The Cost Work Breakdown Structure is also presented as appropriate. Table 12 illustrates the structure of the manufacturing plan section of the Block 50 Manufacturing Task Analysis.

   Subsections 3.2–3.8 contain exploded view diagrams illustrating the assembly processes for their respective areas.

TABLE 12   Structure of the F-16 Manufacturing Plan

Subsection	Title
3.1	General
3.2	Forward Fuselage
3.3	Center Fuselage
3.4	Aft Fuselage
3.5	Wing
3.6	Vertical Fin & Horizontal Stabilizer
3.7	Fuselage Mate, Fuselage Primary, Airframe Mate, Primary Final Assembly and Systems Operations and Field Operations
3.8	Electrical Systems
3.9	Harness Development

3.1 General

This part of the manufacturing plan addresses such items as manufacturing policy, tooling policies, manufacturing support equipment, production plan, component and delivery schedules, and the quality assurance program.

Manufacturing policy. The production plan of the F-16 is designed around the use of conventional materials and structural applications within the scope of current technology. Conventional machines, machine tools, and fabrication equipment are to be used whenever possible.

■ Part fabrication methods are determined on the basis of part complexity and accuracy requirements.

■ The breakdown of major components into smaller subassemblies has been developed in order to establish the most effective sizes, configuration, and method of manufacture.

■ To enhance accessibility and to improve producibility, it is desirable to install the systems items at the component level to the maximum extent possible or practical.

Tooling policies. The basic tooling policy is to realize the lowest possible life-cycle cost. Tooling for existing revisions will be retained for spares provisioning.

■ Work instructions will be prepared to establish the manufacturing sequence for specified product configurations.

■ All work instructions and tool orders will be prepared in strict compliance with the production plan.

■ Tool design drawings will contain a clear, concise, detailed description of the tool and the relation of the tool to the part or assembly being manufactured.

■ Master control tools will be used to fabricate the tools necessary to accomplish their assigned task. They will not be used as production tools.

■ Mock-ups will be used for tool requirements and tool proofing.

■ New tooling shall only be ordered if those from previous versions cannot meet requirements for production.

Manufacturing support equipment. The Manufacturing Support Equipment (MSE) Plan for the Block 50 Program is based on the same concepts used in previous versions of the F-16. Whenever applicable, existing MSE will be used. New configurations of MSE will be developed either by GD or their vendors as needed. MSE is primarily composed of support equipment, F-16 peculiar support equipment, and special test equipment.

Maintenance and calibration is scheduled through the use of Air Force Technical Orders, contract-specific Technical Orders, and GD Standard Practices.

Production plan. The production process sequence follows the one illustrated in Figure 23. Assembly flow times were derived from existing flow times and estimates based on the changes for the Block 50. All engineering releases are done by part number and the related task planning, tool manufacture, material needs, and part manufacture are keyed to the component in which they are installed. Plans are continuously compared and updated against estimates in order to more accurately allocate changes.

(Continued)

Fabrication tasks are scheduled according to the production manufacturing sequence. Raw materials and vendor-supplied items are keyed to the earliest fabrication requirement, as specified by the Production Inventory Control System.

Component and delivery schedules. The Block 50 Component and Delivery Schedules are also presented in the plan. These schedules are determined to accommodate physical and functional mock-ups, pilot production, and rate production. Additionally, the schedules are coordinated with the release of the engineering data.

Quality assurance program. The Block 50 Quality Assurance Program (QAP) is an extension of the on-going F-16 Quality Assurance efforts. The QAP provides control of all the elements necessary to assure air vehicle integrity, conformance to engineering design, and documentation necessary for safety of flight certification. The QAP is broken down into 25 different areas. Some of the representative areas include:

- Measuring and Test Equipment
- Tooling Inspection
- Material and Material Control
- Production Processing and Fabrication
- Training and Certification
- Test Philosophy
- Rework
- Configuration Control

Tool Planning at Bell Helicopter

The Bell Helicopter-Textron facility in Ft. Worth, Texas is where Bell Helicopter designs and produces both their military and commercial aircraft. Recent efforts include producing the V-22 Osprey and competing for the Army's LH Helicopter program.

Both the Osprey and LH designs make extensive use of composite materials. Composites are very difficult to work with, requiring precision and extreme care in handling. Many tools either cannot readily be capable or require extensive modification to handle composites in manufacturing.

Bell Helicopter manufacturing and tooling engineers realized early in the process the need for strategy and planning in any modern helicopter development effort. Additionally, a sound tool plan is a critical component of an effective manufacturing plan.

Strategic objectives

One of the main objectives when competing for projects is to develop cost-effective, lightweight, producible designs. While always a consideration in design, producibility becomes more important as projects lead into the demonstration/validation (dem/val) phase. Here, the design really begins to take shape, and manufacturing and fabrication concerns affect part and tool selection.

For assessing producibility during demonstration/validation, concurrent engineering teams are formed to perform trade-studies. The trade-off-studies are intended to reduce manufacturing risk and evaluate alternative design. Organizations represented on the teams include members with concerns in:

- product assurance
- manufacturing engineering
- design to cost
- supportability
- engineering

Preliminary manufacturing decisions are made with the help of producibility analysis sheets.

Tool planning strategy

As the producibility trade-off studies are being conducted, many issues are raised about tooling, since composites often require specific, custom tools. This allows design, tooling, and manufacturing concerns to be addressed as early in the process as possible. The overall tool plan strategy is illustrated by Figure 25.[122]

The tool plan is a dynamic, living document that is updated and maintained with all of the engineering and manufacturing data as the project moves from demonstration/validation through full-scale development and into production. The tool planning process addresses, at a minimum, six major areas:

- *General Scope*—part requirements, process, tools required
- *Tool Description*—physical, material, tolerances, quality
- *Tool Flow*—illustrations showing the tool in the process, how the tool is to be used
- *Schedule*—trade studies, specific and general tool families, plan documentation
- *Cost*—materials, make-or-buy
- *Inspection*—general, methods, specific, requirements

[122]Courtesy Bell Helicopter-Textron, Ft. Worth, TX.

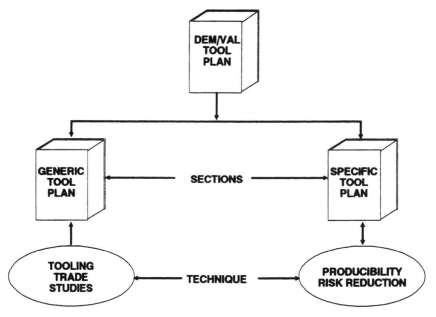

Figure 25   Bell helicopter's tool planning strategy.

Trade study process. Tool requirements originate from part designs process considerations, and producibility analysis reports. Several tool alternatives are defined by tool engineering. The trade study analyses are based upon the requirement inputs and the evaluation criteria. Generally, evaluation criteria are categorized by cost, schedule, and technical merit.

When an alternative is chosen, it is placed into the tool plan, and communicated to the development team. In this way, the manufacturing engineers, tool engineers, and design engineers can make informed, accurate decisions in developing cost-effective, lightweight, producible designs.

The relationship between the development efforts, the manufacturing plan, and the tool plan are summarized in Figure 26.

AT&T's MOS V Qualified Manufacturing Line

AT&T Microelectronics' manufacturing facility in Allentown, PA, demonstrated its ASIC (Application Specific Integrated Circuit) design/MOS V (metal oxide semiconductor V) Wafer Fabrication/JIT Assembly and Test lines capability to the government and was the first company to obtain

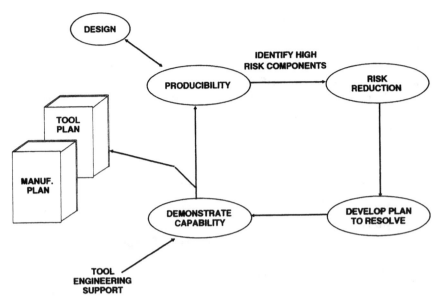

Figure 26   Relationship of tool plan with other activities.

qualified manufacturers list (QML) certification in December 1989. When AT&T demonstrated process stability and predictability on its products, its 1.25 micron CMOS technology was qualified in March 1990. The following provides an overview of the process AT&T went through to obtain QML qualification.[123]

Background

When the DoD used components from vendors without checking for reliability, systems started to fail because of component defects. The failures were due to component failures. This prompted the DoD to require checking every component. As more and more systems used low volume, complex (ASIC, etc.) components, this approach, based on extensive destructive testing of components became too costly. In addition, rigid requirements in the traditional military procurement documents made it difficult for manufacturers to offer their latest innovations/technologies to the military markets. So the DoD changed the rules for component suppliers. The new DoD components specifications

---

[123]Personal communications with W. Vesperman and D. Kane, AT&T Microelectronics, Allentown, PA.

impose statistical process control, and in many cases, maximum failure rates on products.

The Defense Electronics Supply Center (DESC), the DoD agency, began to watch over vendor parts. Rome Laboratory (formerly RADC) and DESC stressed statistical process control and initiated the qualified manufacturer's list (QML) specification for microcircuit production. DoD has created a structured method to certify a manufacturer's facility and thereby validate the reliability of microelectronic components.

> QML specification is only the beginning of a larger DESC thrust, that by 1992, could affect 60,000 electronic components in 56 categories.[124]

In 1986, Rome Air Development Center (RADC) contracted AT&T, GE, and Honeywell to develop *generic qualification* procedures for VHSIC/VLSI devices. The preliminary draft of a specification for QML was issued in 1988.

In 1989, MIL-I-38535, *General Specification for Integrated-Circuit [Microcircuit] Manufacturing,* based on the qualified manufacturing line concept, was issued.

Objectives of QML. Obtaining QML status is a generic way to qualify a manufacturer without extensive end-of-manufacturing qualification testing on each device design. It reduces and replaces the end-of-manufacturing testing with in-line monitoring and testing using SPC. It shifts the focus from device-level to process-level.

The foundation for this approach is a TQM approach within the manufacturing environment. It enables a manufacturer to apply for the Malcolm Baldrige National Quality Award within five years of the initial request for the QML status.[125] QML approach certifies processes rather than individual parts.

The key objectives of the QML are:

- built-in quality

- training and motivation of all employees

- continual improvement

---

[124]Keller, John. "Defense Electronics Supply Center: Lending the QML Approach to All Electronic Spare Parts." *Military & Aerospace Electronics.* vol. 1 (7), July 1990, p. 49.

[125]Spurgeon, Susan P., Marcinko, Frank, Mengele, Martin J., Lyman, Richard C. "QPL or QML—A Quality Trilogy Approach." *1990—ASQC Quality Congress Transactions— San Francisco.* 1990, pp. 189-193.

The two steps in achieving QML status are: certification and qualification.

Certification requirements. The certification requirements are:

- controlled processes
- continued improvement
- quality-management approach
- self-audit with quality enhancement
- technical-review board (TRB) acting as corporate conscience
- verified reliability of process

Qualification requirements. The qualification requirements are:

- demonstration of the stability and predictability of the manufacturing line (for example, two actual microcircuits and the standard evaluation circuit [a line-monitoring vehicle] must be run on the QML-certified line).
- comparison of qualification, yield, reliability, and failure-mode analysis to data on those two products
- demonstration of a good correlation and satisfactory reliability

QML as a business strategy

AT&T found the QML concept to be a good business practice because it ensures cost-effective programs and operations. AT&T believed that obtaining QML qualification is consistent with various quality initiatives in practice in-house. In addition, AT&T was a part of the team that drafted the generic qualification procedures for integrated circuits.

AT&T decided to follow the QML concept to develop a single manufacturing process for both commercial and military systems. AT&T realized the following benefits of QML qualification and incorporated QML in its business strategy:

- improved quality
- improved efficiency of operations
- long-term cost reductions
- improved customer satisfaction

Figure 27 shows AT&T's conversion of customer requirements under QML.

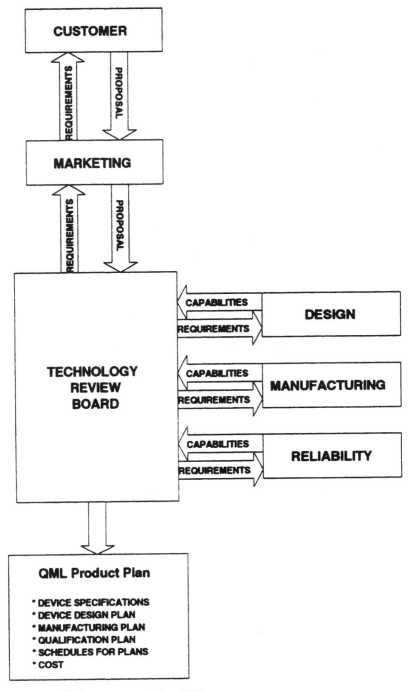

Figure 27   AT&T's process to achieve QML status.

Planning for QML qualification

AT&T started QML planning by:

- identifying and translating customer and quality-assurance requirements given in MIL-I-38535, *General Specification for Integrated Circuit Manufacturing,* into internal systems and documents
- doing a self audit against MIL-I-38535 specifications to ensure that all needs were met
- identifying the metrics to evaluate the process
- understanding the role of SPC, process monitors

AT&T developed a Total Quality Management (TQM) plan to include the key aspects of QML: Technology-Review Board, process flow, design process, line-monitoring program, reliability-monitoring program, quality-improvement plan, yield-improvement program, change-control program, device-qualification plan, field-monitoring program, self-audit program, and failure-mode-analysis program.

Technology-review board. AT&T formed a technology-review board (TRB) which is a technical management organization to oversee the technical, quality, and reliability issues for QML products. The TRB has representatives from design, wafer fab, assembly, test, product engineering, quality control, qualification, reliability, and marketing organizations. A quality manager chairs the TRB and has absolute veto power.

The TRB is actively involved in all aspects of certification, qualification, and manufacturing. The TRB generates reports on its activities to the government.

Process flow. AT&T defined the total process flow for its MOS QML line (integrated-circuit-line) to include: design centers, mask manufacturing, wafer fabrication line, assembly line, and test center. AT&T ensured that all elements of the process flow and interfaces among those elements met QML requirements.

Design process. AT&T's ASIC Design Center complied with QML requirements and concepts by ensuring and documenting the necessary controls of the basic elements of the design process:

- the technology database (models and design rules)
- an integrated CAD/CAE system
- standardized design practices and procedures
- all interfaces between the customer and manufacturing

Line-monitoring program. To monitor the processes in-line and maintain a controlled line during production, AT&T used a line-monitoring program that included:

- a statistical process control plan
- a standard evaluation circuit
- a technology-characterization vehicle
- process monitors
- process zone monitors

Statistical process control plan. AT&T developed an SPC plan to monitor the process for unwanted changes and to maintain stability. The plan included capability to:

- identify critical nodes and establish control charts
- do process-capability studies on all critical nodes
- calculate process-capability-index values for all critical nodes
- continuously cause improvement at all critical nodes

Standard evaluation circuit (SEC). AT&T designed a functional microcircuit called SEC to monitor and control the manufacturing process. SEC serves as a well-characterized, easily diagnosable standard product. It is used to exercise the existing design documentation, software tools, and performance simulations. It is also used to demonstrate design and manufacturing reliability.

Technology characterization vehicle (TCV). The TCV is used to characterize a technology's susceptibility to intrinsic reliability failure mechanisms. AT&T dedicated the majority of the Tester Reliability Yield Component (TRYC) area to yield testers, which monitor the defect densities of various process zones after the full process sequence.

Process (parametric) monitors. AT&T uses a family of test structures called process monitors to measure electrical characteristics of each wafer type containing the product chips. The process monitors are positioned to determine the uniformity across the wafer. The use of process monitors enables manufacturing to select an individual wafer or a wafer lot based on whether it met the pass or fail criteria.

Process zone monitors. AT&T implemented process zone monitors to monitor defect density for particular processing segments of the wafer fabrication. The process zone monitors are used to:

- evaluate changes to equipment
- evaluate process changes
- enhance process checks
- improve the yield by reducing defect density

Reliability-monitoring program. AT&T set up a reliability-monitoring program to define inspection procedures to ensure that the device and lot quality requirements are met. The reliability monitoring program addresses all aspects of MIL-STD-883C Method 5005, *Qualification and Quality Conformance Procedures.* Data were generated at an increased frequency with increased sample sizes under this program. SEC data are correlated to data from other products in the same technology. The program replaces some end-of-line tests with in-line rigorous process controls when appropriate.

AT&T set up a reliability-review board (RRB) consisting of experts. The RRB reports to the TRB. The RRB evaluates the reliability data, and corrective actions and decided the product disposition procedures.

Quality-improvement plan. AT&T implemented a quality-improvement plan with a strong focus on quality and customer satisfaction. The quality-improvement plan included setting up quality-improvement teams who defined the yield and quality requirements. The plan also included training and sensitization of all employees to various aspects of quality.

Yield-improvement plan. AT&T developed a yield-improvement plan to remove yield-limiting defects. The objective of this plan is to improve quality, reliability, and customer satisfaction. This plan included use of SPC, SEC, TCV, parametric and zone monitors to drive yields. The plan also included a yield model to identify defect-density structures and predict yields.

Change-control program. AT&T set up a qualification-review board (QRB) comprising reliability experts from all areas of the process flow. The QRB developed the change-control program which required all changes to be fully documented and approved by quality, manufacturing, and R&D managers. The QRB assesses all changes for their possible impact on product reliability. The QRB defines the following test requirements to prove-in changes:

All changes should be reported to the TRB.

All major changes require TRB approval prior to implementation.

Major changes require customer approval prior to implementation.

Device-qualification plan. The QRB developed the device-qualification plan using MIL-STD-883C, Method 5005. Documented guidelines helped the QRB to group devices into processing and packaging families whose reliability characteristics are likely to be identical. The QRB performs tests for all known failure mechanisms and stores all reliability data (from qualification, monitoring, etc.) in a database.

Field-monitor program. AT&T Customer Technical Support Center manages a highly active field-monitor program that:

- receives and acknowledges complaint/query
- coordinates and tracks timely solution
- documents results of failure-mode analysis (FMA) for the customer
- maintains database of all device issues
- disseminates customer concerns to management and TRB
- advises customers of quality concerns and product changes
- recalls devices for upgrading and screening
- coordinates visits to customer locations and hosting customer visits

Self-audit program. AT&T has a very active self-audit program to verify the adequacy and implementation of total-quality program. This program included self-audit that:

- documents findings (deficiency notices)
- tracks deficiency notices to resolution
- verifies corrective actions
- reports monthly to TRB and higher management

Lessons learned

The following summarizes the valuable lessons the AT&T team learned during QML certification and qualification:

- Management support is a must for the success of the program.
- Educating and sensitizing all employees to quality makes the transition smooth.
- Up-front involvement of the manufacturer, DESC/RADC, and validation teams simplify the process.
- QML makes good business sense and can be used for commercial applications also.

Future direction

AT&T is using QML concepts and disciplines to develop the manufacturing practice for both commercial and military applications. The goal is to economically improve manufacturability, quality, and reliability. AT&T plans to have a uniform manufacturing process for all commercial and military products.

Paperless Assembly Line for the F/A-18 Hornet

In a plane, the computer must work or the plane can't fly. At Northrop, the computer must work or the planes can't get built.[126]

Northrop Aircraft Division (NAD) production plant in Hawthorne, CA uses a computer to run its facilities. NAD assembles the F/A-18 Hornet fighter, an aircraft used by the U.S. Navy and Marine Corps in this plant. The assembly line combines the elements that make up the center and aft fuselage sections, the twin vertical stabilizers, and the associated subsystems.

Background

In 1978 Northrop began building the F/A-18 in its plants in Hawthorne and nearby El Segundo. Between 1978 and the first few months of 1989 it produced 842 assemblies, or "shipsets," as they are known in the aircraft field. Each shipset requires 16,295 sheets of paper—schedules, part lists, tooling, inspections, nonconformance data. During the Paper Age (i.e., 1978-89) this plant produced more than 13.45 million sheets of paper.[127]

Integrated planning and control for assembly

In February 1989, NAD converted its F/A-18 final assembly to an entirely paperless manufacturing shop-floor control system. NAD developed the computerized system that manages the production line, Integrated Planning and Control for Assembly (IMPCA), in-house. The IMPCA system was critical to the production process and was on a fault-tolerant Tandem computer system.[128]

---

[126]Vasilash, Gary S. "Manufacturing by Wire." *Production,* June 1990, p. 85.

[127]Vasilash, Gary S. "Manufacturing by Wire." *Production,* June 1990, pp. 85-89.

[128]Best Manufacturing Practices (BMP) Review Team. *Best Manufacturing Practices, Report of Survey Conducted at NAD,* Hawthorne, CA. Washington, DC: Office of the Assistant Secretary of the Navy (Shipbuilding and Logistics), March 1989, pp. 14-16.

By converting to the IMPCA system, NAD eliminated 400,000 pieces of paper at 104 stations in 14 cost centers on the line. The projected cost savings through the F/A-18 contract was $20 million. The IMPCA system performed:

- work planning, instruction, and dispatching
- resource allocation
- work performance monitoring and evaluation
- maintenance of electronic records

Conversion plan.   The conversion plan was critical to the success of the IMPCA program. Four key aspects of the plan were:

- informing and educating affected employees
- developing prototypes and doing simulations for selected centers
- installing and transferring to IMPCA without disrupting the line and the delivery schedule
- validating the new system as functionally equivalent to the previous paper-driven one

During the conversion to the automated system, 15 shipsets were produced using both the paper-driven and the electronic methods. The first five of the 15 shipsets were used to introduce the staff to the system and were not a part of the validation process. The last ten were used by the Air Force Plant Representative Office for in-depth validation of the new system. When minor discrepancies were found during validation, they were equally likely to be in the paper-driven system or the new one.

Other key aspects of NAD's approach to the conversion plan that contributed to success are:

- simulation of three production work centers to better explain shop-floor activities
- approval of changes only by a change-control team made up of users, data-processing staff, manufacturing-initiatives-process staff, and project engineers
- password and electronic-stamp security built in the system that limits document ownership to only one user at a given time

Training.   NAD recognized that perhaps the biggest obstacle was employee acceptance. Fear of the system was overcome through briefings and training programs that began two years before the conversion.

The system included user-developed guide screens that dealt in specific terms and methods familiar to the users. During the final stages of the education process, experienced users in each center conducted the hands-on training for their fellow employees.

IMPCA capabilities. At the center of the IMPCA system is an on-line transaction processing fault-tolerant computer. The system needs to be up for production all the time except when it is under maintenance for 20 hours per week. The IMPCA system enables in real-time to:

- determine quality trending
- discover backlogs or bottlenecks
- schedule quality assurance
- make engineering changes
- schedule operations

Summary

The system has improved the overall quality of the production process by providing real-time access to data. Approved users can access data at any terminal. The system's monitoring capabilities assure that tasks are done properly. New data gathered is available promptly to adjust time standards. Up-to-the minute status information, which indicates the progress of jobs, is available on-line.

In the post-Paper Age, each plane is fully documented by 36 (4 in. by 6 in.) microfiche cards. Gary L. Lampkins, Manager, F/A-18 Assembly Operations IMPCA claims that it is the only paperless aircraft factory in the world.[129]

With each succeeding shipset produced in a paperless mode, NAD has become more confident about the IMPCA system and paperless manufacturing. Migration plans are being generated for other NAD programs, such as the 747 and B-2.

Air Force Lessons Learned Program

The Air Force Acquisition Logistics Division (ALD) is intended to bridge the gap between the acquisition and logistics communities. Its main goal is to help improve reliability and supportability of new weapons systems coming into the Air Force inventory, while lowering the cost of ownership for those systems. One method that is employed

---

[129]Vasilash, Gary S. "Manufacturing by Wire." *Production,* June 1990, pp. 85-89.

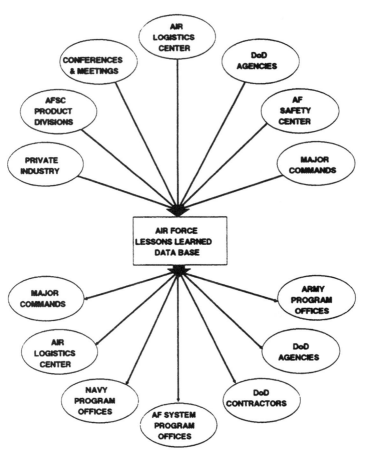

Figure 28   Air Force lessons learned program.

by ALD is the recording and sharing of the experiences of past programs; the Lessons Learned Program.

The Directorate of Lessons Learned and Systems Support (ALD/LSL) maintains a database of lessons learned. Since 1977, the database has collected in excess of 2000 validated lessons learned. Figure 28 illustrates how the ALD/LSL collects and distributes the lessons within the DoD community.

The lessons are broken down into two areas within the database. Management lessons address program decisions and actions such as planning, configuration management, and program control. Technical lessons relate to such things as systems, equipment, design factors, and performance factors. A collection of lessons learned abstracts, covering

47 different areas, is available from ALD/LSL. Some of the abstracts related to the issues in this guide are cited below:[130]

Vendor quality and configuration control. Lack of emphasis on and enforcement of certain contract provisions concerning quality and configuration control by the system program office, contracting office, and contractor leads to deficient quality assurance programs, unauthorized actions, and deficient supplies.

Disposition of special tooling from production contracts. Improper or late identification of special tooling used in production may result in inappropriate decisions regarding future use.

Manufacturing process. Failure to ensure that manufacturing processes are followed in detail can allow defective equipment to be delivered to the government.

Contractor quality programs. A contractor's quality programs should be evaluated in addition to plant capability prior to contract award.

Manufacturing technology/modernization. If producibility and productivity improvements are not considered throughout the life of a production process, cost savings and schedule improvement opportunities will be missed.

"Historical data" or "averages" used by contractors in cost and pricing data. Utilizing cost averaging based on "historical data" that may not reflect current practices may result in overpricing of new or follow-on contracts.

Manufacturing and finishing processes. Changes to manufacturing and finishing processes that seem small or insignificant can have a large impact on finished product durability.

Manufacturing, contractor, and subcontractor capability. Inadequate assessment of contractor/sub-contractor manufacturing capability, business practices, and financial health prior to contract award may result in increased costs and/or schedule slippage.

Tool and hardware control by manufacturers and maintenance contractors. Some manufacturers and maintenance contractors are not controlling tools and hardware and are not ensuring that manufacturing/maintenance debris is removed after task completion.

---

[130]Department of the Air Force, Lessons Learned Program, *Abstract of Lessons Learned,* Wright-Patterson AFB, OH: ALD/LSL, October 1, 1989.

Quality assurance of production tooling. Government quality assurance personnel must ensure contractors have complete control of production tooling used in new procurement in order to preclude the manufacturing of nonconforming material.

Maintenance plan. Inconsistent policy on maintenance planning will have an adverse effect on weapon systems acquisition.

Training plans. Untimely or incomplete training plan development can result in insufficient lead time to support milestone decisions and acquire needed resources.

Management of special tooling/special test equipment. Lack of sufficient information often results in poor decisions whether to dispose of or retain ST/STE.

4

Summary

Manufacturing-related efforts span the entire life cycle of a product. Strategic manufacturing planning emphasizes the determination, planning, and implementation of activities that can transform manufacturing into a competitive advantage.

> The purpose of manufacturing is to serve the company—to meet its needs for survival, profit, and growth. Manufacturing is part of the strategic concept that relates a company's strengths and resources to opportunities in the market.[131]

Manufacturing as a strategic necessity is not a new topic. Also, continuous improvement is not a new topic. As far back as the 1920s, people like Walter Shewhart of AT&T Bell Laboratories, advocated understanding and controlling processes in order to detect problems early and turn them into opportunities for improvement instead of costly mistakes.

Strategic manufacturing planning makes good business sense. As the AT&T case emphasized, management support, education, and up-front involvement of everyone are essential to success. For too long, the reaction to implementing continuous quality improvement programs was similar to this:

> When we were told that a new program was being done for our mutual benefit, we had the same reaction I did when my mother told a younger-me to eat my Brussels sprouts.[132]

---

[131]Skinner, Wickham. "Manufacturing—Missing Link in Corporate Strategy," *Harvard Business Review,* May/June 1969, p. 140.

[132]Huber, Robert F. "Quality: Survival, Not a Snow Job," *Production,* February 1991, vol. 103(2), p. 9.

This attitude must change if a company expects to survive in a competitive marketplace. Manufacturing must be well understood and integrated into the entire set of corporate goals and activities. This reference guide has attempted to illustrate the process of strategic manufacturing planning, and to alert the reader to risks inherent in the process and emphasize the keys to success. The process itself spans a complex undertaking that requires understanding, foresight, teamwork, and commitment. The Application chapter highlights case studies that reflect how some firms have chosen to implement many of the points presented in the Procedures chapter. Table 13 summarizes the steps of the process, as described in Procedures.

TABLE 13   Summary of Strategic Manufacturing Planning Procedures

Step 1: Develop Manufacturing Strategy

Procedures	Supporting Activities
Understand the role of manufacturing strategy	Understand the definition of manufacturing strategy
	Understand the relationship of manufacturing strategy with other strategies
	Examine and determine strategic decision areas
	Assess state of manufacturing's strategic effectiveness
Evaluate manufacturing choices	Develop key capabilities
	Develop the strategic plan
	Balance key capability and strategic plan development
Identify key performance measures	Establish an effective manufacturing performance measurement system
Identify opportunities	Compare with strategic objectives
	Compare with strengths and weaknesses
Evaluate manufacturing strategy	Determine appropriateness of strategy in terms of completeness and competitiveness
Implement manufacturing strategy	Establish and promote education, teamwork, management support, and consensus

Step 2: Assess Present Capabilities and Desired Improvements

Assess Present Capabilities	Determine "as-is" and "to-be" states
	Identify strategic strengths and weaknesses (e.g., producibility, processes, tooling, facilities, human resources, training)
	Review lessons learned

**TABLE 13   Summary of Strategic Manufacturing Planning
Procedures (*Continued*)**

Step 2: Assess Present Capabilities and Desired Improvements (*cont'd*)

Procedures	Supporting Activities
Assess Present Capabilities (*cont'd*)	Understand role in marketplace
	Establish performance measures
	Determine constraints
	Evaluate cost-accounting methods
Compare Objectives	Understand current initiatives and their status
Assess desired improvements	Establish priorities and compare with manufacturing strategy
	Understand cost-based decision practices
	Consider activity-based costing
Identify Programs for Improvement	Develop improvement plans
	Do make-or-buy analysis
	Determine funding and resource requirements
Do external analysis	Examine competitive benchmarking data
	Consider initiatives such as Industrial Modernization Incentives Program (IMIP) and Manufacturing Technology (MANTECH)
	Identify potential partners
	Use technology transfer
Decide whether to pursue opportunity	Assess match between project and strategies

Step 3: Develop Manufacturing Plan

Understand the manufacturing planning process	Start manufacturing planning activities as early as possible
	Maintain connectivity with the rest of the project
	Understand the role of manufacturing management
	Define the scope of manufacturing management and consider issues involved
	Establish configuration management and change-control procedures
Define the guidelines for subcontractor relationship	Examine the subcontractor's quality-improvement program
Understand issues in the manufacturing process	Consider part fabrication and assembly, tolerances, material handling, and test and inspection

TABLE 13   Summary of Strategic Manufacturing Planning Procedures (*Continued*)

Step 3: Develop Manufacturing Plan (*cont'd*)

Procedures	Supporting Activities
Develop manufacturing plan	Ensure that the plan matures concurrently with the project
	Ensure plan completeness
Develop Tool Plan	Understand the role of tooling, special tooling, and special test equipment
	Evaluate tooling needs
	Evaluate special test equipment needs
	Compare tooling/test equipment needs with strategic constraints
Develop Manufacturing Process Qualification Plan	Understand the importance of process qualification
	Characterize the process
	Standardize the process
	Evaluate the process
	Consider benefits of QML activities
Develop Factory Improvements Plan	Understand the concept of the factory of the future
	Study the role of computer-integrated manufacturing
	Consider maintenance issues
	Consider material handling, flow, and inventory issues
	Study plant layout issues
	Consider plant operations issues
	Understand JIT
Develop Operations Plan	Assess the role of the operations plan
	Do process planning
	Coordinate and communicate work instructions
	Include plant maintenance issues

Step 4: Monitor, Control, and Improve

Implement process-oriented management	Select methodology to manage and improve process quality
	Study the principles of the methodology
	Establish process ownership
	Assess the process
	Understand the process
	Monitor the process
	Collect and analyze data
	Select appropriate analysis tools

TABLE 13   Summary of Strategic Manufacturing Planning Procedures (*Continued*)

Step 4: Monitor, Control, and Improve (*cont'd*)

Procedures	Supporting Activities
Select opportunity for improvement	Review problem areas
	Rank improvement opportunities
	Gain consensus on opportunity selected
Improve	Organize a process improvement team and develop an action plan
	Determine root cause of the problem
	Test and implement the solution
	Hold process gains and follow through
	Perform periodic process review

Chapter

5

References

AT&T Quality Steering Committee. *Process Quality Management & Improvement Guidelines.* Indianapolis, IN: AT&T Customer Information Center, January 1989. Describes a structured method for managing and improving quality of service and administrative functions. Describes proven tools and techniques to support process quality management and improvement activities.

Acker, David D. and Young, Sammie G. LTC, USA. *Defense Manufacturing Management Guide for Program Managers.* Ft. Belvoir, VA: Defense Systems Management College. 3rd Ed., April 1989. Describes effective manufacturing management methods used in defense systems acquisition programs. Describes manufacturing strategy, manufacturing planning and scheduling.

Aerospace Industries Association. *Restoring Old Glory: A Strategy for Industrial Renaissance.* Washington, DC: Aerospace Industries Association of America, 1988. Describes IMIP and its benefits.

Barlas, Stephen, "Army Takes Flak Over Flexible Manufacturing," *Managing Automation,* vol. 5(12), December 1990, p. 20. Notes some conclusions from a recent Manufacturing Studies Board report on the US Army's industrial mobilization planning efforts.

Benassi, Frank. "Honeywell Integrates Building and Process Controls in Factories," *Managing Automation,* vol. 6(3), March 1991, pp. 38-40. Describes *the integrated manufacturing facility* developed by Honeywell.

Benassi, Frank. "Maintenance Management: Manufacturing's Final Frontier," *Managing Automation,* vol. 6(3), March 1991, pp. 34-36. Describes the concept of total productive maintenance and the importance of predictive and diagnostic maintenance in cycle reduction and just-in-time manufacturing.

Bergstrom, Robin P. "Some Things Just Don't Translate Well," *Production,* vol. 102(12), December 1990, pp. 54-57. Discusses with examples the importance of material handling systems in supporting the manufacturing strategy.

Bhaskaran, Kumar. "Process Plan Selection," *International Journal of Production Research,* vol. 28(8), 1990, pp. 1527-1539. Presents a model which highlights key issues and provides an analytical basis for selecting process plans.

Braham, James. "What Price Stealth?", *Machine Design,* vol. 63(4), February 21, 1991, pp. 26-28. Highlights the problems encountered in the A-12 development program, and presents some findings from the A-12 Administrative Inquiry.

Bryce, G. Rex. "Quality Management Theories and Their Application," *Quality,* February 1991, p. 24.

Burgess, Lisa. "Thomas: Pushing the Pentagon Toward QML," *Military and Aerospace Electronics,* February, 1990, pp. 39-40. Describes Robert Thomas's support of TQM and Qualified Manufacturing Line concept for "living" or "generic" specifications.

Butler, Michael P. "Facility and Capacity Planning Using Sales Forecasting by Today's Industrial Engineer," *Industrial Engineering,* June 1990, pp. 52-55. Describes how to develop a facilities strategy using sales forecasting as a tool.

Calkins, J. "Manufacturing Strategy: A Greenfield Approach," paper presented at the 4th Annual Best Manufacturing Workshop, September 12, 1990, Scottsdale AZ. Describes TRW's manufacturing strategy, implementation, and status of the MEAD's facility in Rancho Carmel, CA.

Cooper, Robin. "Elements of ABC", in *Emerging Practices in Cost Management*, Ed. Barry Brinker, NY: Warren, Gorham, & Lamont, 1990, pp. 3–23. Describes activity-based costing and compares it with conventional cost-accounting systems.

Cooper, Robin. "Introduction", in *Emerging Practices in Cost Management*, Ed. Barry Brinker, NY: Warren, Gorham, & Lamont, 1990, pp. xiii–xvii. Describes the contents of the book and defines cost management. Describes activity-based costing and other cost management techniques.

Cranfill, Sidney M., "Seven-Task Manufacturing Improvement Program," *Manufacturing Systems*, vol. 9(1), January 1991, pp. 35-40. Describes the seven steps in manufacturing improvement programs.

Crosby, David C. "How to Succeed in SPC," *Quality in Manufacturing*, vol. 2(2), March/April 1991, pp. 36–37. Describes seven fatal mistakes to avoid when implementing SPC. Provides a 20-point system for evaluating the quality management system.

DI-MISC-80074. *Manufacturing Plan*, February 1985. Identifies the contractor's overall system and detailed factors necessary to achieve an effective, efficient manufacturing program.

Danzyger, Howard. "Strategic Manufacturing Plan? Your Competitiveness Depends on It." *Industrial Engineering*, vol. 22(2), February 1990, vol. 22(2), pp. 19-23. Describes strategic plans and their benefits and how to transition from the current to future environments.

Deloitte & Touche. "Issues in Competitive Manufacturing." 1987. A series of articles describing topics such as JIT, strategic cost management, electronic data interchange, and design and manufacturing restructuring.

Deming, W. Edwards. *Out of the Crisis*. Cambridge, MA: MIT Center for Advance Study, 1982. Illustrates the transformation with examples the style of American management required for survival. Describes what Deming believes American managers have been doing wrong and what they must do.

Department of Defense. *Transition from Development to Production*. DoD 4245.7-M, September 1985. Describes techniques for avoiding technical risks in 47 key areas or templates in funding, design, test, production, facilities, logistics, management, and transition plan.

Department of Defense. *Industrial Modernization Incentives Program (IMIP)*. DoD 5000.44-G, August 1986. Provides guidance in the use of IMIP an an acquisition tool to enhance productivity, improve quality, and reduce acquisition costs.

Department of the Air Force, Lessons Learned Program, *Abstract of Lessons Learned*, Wright-Patterson AFB, OH: ALD/LSL, October 1, 1989. Presents short abstracts from the Lessons Learned Program database, sorted by 47 different categories.

Department of the Navy—Best Manufacturing Practices Program, *Producibility Measurement for DoD Contracts*, 1990. Describes what is necessary to ensure that producibility is addressed correctly. Also presents a set of producibility measurement worksheets.

Department of the Navy. *Best Practices: How to Avoid Surprises in the World's Most Complicated Technical Process*. NAVSO P-6071, March 1986. Describes how to avoid traps and risks by implementing best practices for 48 areas or templates including manufacturing strategy, manufacturing plan, qualify manufacturing plan, factory improvements plan, operations plan, and tool plan.

Design to Reduce Technical Risk. Describes industry and government best practices for design policy, the overall design process, and design analyses; design release and configuration control; planning and implementing computer-assisted technology; applying design reviews; and successful implementaion of the Front-End Process.

The Determinant Factors for a Successful MRPII Implementation, Saratoga Springs, NY: Business Education Associates, 1987. Describes the research on MRP II conducted at the SUNY Albany School of Business.

Donnell, Augustus and Dellinger, Margaret. *Analyzing Business Process Data: The Looking Glass.* Indianapolis, IN: AT&T Customer Information Center, 1990. Describes basic statistical quality control (SQC) for business operations.

Doshi, Bharat T. and Krupka, Dan C. "Integration of Planning and Execution Operations: Theory and Concepts," *AT&T Technical Journal,* vol. 69(4), July/August 1990, pp. 90-98. Describes material requirements planning (MRP) systems and just-in-time (JIT) techniques and how they coexist in a system.

Dumas, Roland A., Cushing, Nancy and Laughlin, Carol. "Making Quality Theories Workable," *Training & Development Journal,* February 1987, pp. 30-35.

Dwyer, James P., Major, USAF. *The Manufacturing Management Officer's Handbook,* Air Command and Staff College Report 85-0730, Maxwell AFB, AL : ACSC/EDCC, April 1985. Describes the Systems Program Office (SPO) responsibilities which the Air Force Manufacturing Officer should be aware of.

Eighth Annual Computer Survey. "Data Collection: The Key to Efficiency," *Modern Materials Handling,* November 1990, p. 66.

Fife, William J. Jr. "The Automation Imperative," *Assembly,* 1990 Buyer's Guide Issue, July 1990, pp. 216. Describes how automation offers the key to industrial survival.

Gardner, Fred. "Hold Down Ballooning Costs and Boost Quality," *Electronic Purchasing,* June 1988, p.57.

Goodrich, Thomas. "JIT & MRP CAN Work Together," *Automation,* vol. 36(4), April 1989, pp. 46-48. Dispels the misconception that MRP II and JIT systems are not compatible.

Gorte, Julie Fox. "Competing in Manufacturing: What Industry Can Do," *Manufacturing Systems,* January 1991, vol. 9(1), pp. 63-64. Describes the second part of the study of the Office of Technology Assessment (OTA) of the U.S. Congress titled "Making Things Better: Competing in Manufacturing." Compares Japanese and U.S. manufacturing methods and identifies specific actions to improve productivity in manufacturing enterprises.

Gould, Lawrence. "Putting a CAPP on CIM," *Managing Automation,* vol. 5(8), August 1990, pp. 17-19. Describes the two major types of CAPP systems, their differences, benefits, uses, and implementation considerations.

Grauf, William M. "Lead Time Management, The Missing Link Between MRP II and JIT," *P&IM Review with APICS News,* August 1990. Discusses management of lead time as fundamental in migrating to JIT from a MRP II environment.

Hammes, Sara and Tricia Welsh. "The Top 25 Contractors," *Fortune,* vol. 123(4), February 25, 1991, pp. 68-69. Provides a list of the top 25 contractors by contract dollars and their major weapons systems.

Harmon, Roy L. and Peterson, Larry D. *Reinventing the Factory.* NY: Free Press, 1990. Describes topics such as focused factory organization and paperless factory.

Havatny, Josef. "Dreams, Nightmares, and Reality," *Computers in Industry,* vol. 4(2), 1983, pp. 109-114. Surveys the history of computer-controlled manufacturing systems over the last thirty years.

Hazeltine, Frank W. "The Key to Successful Implementations," *P&IM Review with APICS News,* November 1990, vol. 10(1), pp. 40-41, 45. Describes how to successfully implement new initiatives.

Huber, Robert F. "Quality: Survival, Not a Snow Job," *Production,* February 1991, vol. 103(2), p. 9. Presents a perspective on quality.

Johansson, Henry J. "Preparing for Accounting Changes," *Management Accounting,* July 1990, pp. 37-41. Discusses the fact that while manufacturing technologies and philosophies have changed, accounting and managerial control systems have not.

Juran, Joseph M., *Managerial Breakthrough.* NY: McGraw-Hill, 1964. Provides a complete understanding of the universal consequences which enable managers to perform better.

Kaeli, James K. "A Company-wide Perspective to Identify, Evaluate, and Rank the Potential for CIM," *Industrial Engineering,* vol. 22 (7), July 1990, pp. 23-26. Presents a systematic approach to identify, evaluate, and rank CIM potential across the enterprise.

Keller, John. "Defense Electronics Supply Center: Lending the QML Approach to All Electronic Spare Parts." *Military & Aerospace Electronics,* vol. 1 (7), July 1990, pp. 49-50. Describes how DESC's support of the QML for electronic parts.

Keller, John. "Industry Wrestles with the JAN to QML Transition," *Military & Aerospace Electronics,* vol. 1 (7), July 1990, pp. 13-14. Describes the payoffs of the move from QPL to QML. Discusses the possibility of acceptance QML by Class-S community.

Keys, David E. "Limitations of Cost Accounting in an Automated Factory," *Computers in Mechanical Engineering,* July/August 1987, pp. 26-29. Discusses some limitations of cost accounting that have become more apparent with increasing automation.

Keyser, Jack. "Manufacturing Process Control," in *Microelectronic Reliability,* ed. Edward B. Hakim, Norwood, MA: Artech House, 1989. Presents a disciplined approach to understanding process variation and process improvement. Discusses process characterization, process capability studies, process optimization, process and product control and monitoring, and process improvement.

Koelsch, James R. "Manufacturing Strategy: The Secret Weapon," *Machine and Tool BLUE BOOK,* vol. 84(9), September 1989, pp. 43-48. Describes the evolution of manufacturing from 1960s to 1990s and the role of strategic planning.

Koska, Detlef K. and Romano, Joseph D. *Countdown to the Future: The Manufacturing Engineer in the 21st Century,* Profile 21 Executive Summary, Dearborn, MI: SME, 1988. Provides a broad perspective of the future environment, tools, and the roles of manufacturing engineers.

Lubben, Richard T. *Just-In-Time Manufacturing,* NY: McGraw-Hill, 1988. Provides an overview of implementing JIT manufacturing.

Manufacturing Studies Board. *Toward a New Era in U.S. Manufacturing: The need for a National Vision.* Washington, DC: National Academy Press, 1986. Describes recent trends in manufacturing and the changes necessary to maintain the competitiveness of US manufacturing.

McClelland, M.N. "An Action Plan for Manufacturing Improvement," *Manufacturing Systems,* October 1990, vol. 8(10), pp. 60,63. Discusses process planning and analysis as helpful in improving cycle efficiency and standardizing manufacturing methods.

MIL-HDBK-50A, *Evaluation of a Contractor's Quality Program,* June 26, 1990. Establishes the means to evaluate a contractor's quality program.

MIL-M-38510, *General Specification for Microcircuits,* February 12, 1988. Establishes the general requirements for monolithic, multichip, and hybrid microcircuits and the quality and quality assurance requirements that must be met in their acquisition.

MIL-M-38535, *General Specification for Integrated Circuits Manufacturing,* December 18, 1989. Establishes the general requirements for integrated circuits or microcircuits and the quality and quality assurance requirements that must be met in their acquisition.

MIL-STD-1528A. *Manufacturing Management Program,* September 1986. Prescribes manufacturing management objectives and requirements that must be met by the contractor's manufacturing system on any contract that incorporates this standard.

MIL-STD-883C, Method 5005. *Qualification and Quality Conformance Requirements,* February 12, 1988. Establishes the general requirements for monolithic, multichip, and hybrid microcircuits and the quality and quality assurance requirements that must be met in their acquisition.

McDougall, Duncan C. *Effective Manufacturing Performance Measurement Systems: How to Tell When You've Found One.* Boston University School of Management, February 1988.

McNair, C. J. "Interdependence and Control: Traditional vs. Activity-Based Responsibility Accounting," in *Emerging Practices in Cost Management,* Ed. Barry Brinker, NY: Warren, Gorham, & Lamont, 1990, pp. 421–430. Describes how organizational interdependence changes the assumptions underlying traditional responsibility accounting. Describes how activity-based costing matches the demands of interdependence and continuous learning in an organization.

Metz, Sandy. "Making Manufacturing Better, Not Just Faster," *Managing Automation,* vol. 5(8), August 1990, pp. 22-24.

Miller, Jeffrey G., Hayslip, Warren. "Implementing Manufacturing Strategic Planning," *Planning Review,* July/August 1989, pp. 26-27, 48. Explains why and how the "global" manufacturing strategy should be integrated into the corporate strategy. Describes the development and implementation of a manufacturing strategy.

Moody, Patricia E. *Strategic Manufacturing,* Homewood, IL: Dow Jones-Irwin, 1990. Describes strategic manufacturing planning process, strategies, tactics, and dynamics of change.

Nazaruk, Pam. "Test Process not Product, Orders the Pentagon," *Electronic Business,* October 15, 1990, vol. 16(19), pp. 163-164. Describes the Pentagon's four-pronged TQM strategy.

Nevins, James L. and Whitney, Daniel E. *Concurrent Design of Products and Processes,* NY: McGraw-Hill, 1989. Discusses a strategy for the next generation of manufacturing by focusing on approaching concurrent design.

O'Guin, Michael C., "Activity-Based Costing: Unlocking Our Competitive Edge," *Manufacturing Systems,* vol. 8(12), December 1990, pp. 35-40. Describes how ABC can be used to achieve a competitive advantage.

Pearce II, John A., Robinson, Jr., Richard B. *Strategic Management.* Homewood, IL: Richard D. Irwin, 1982. Describes the critical business skills of planning and managing strategic activities that constitute the strategic management process. Illustrates the interdependence of various parts of the strategic management process through examples.

Pfeiffer, J. William. *Strategic Planning: Selected Readings.* San Diego: University Associates, 1986. Presents selected readings which cover such topics as the focused factory organization, process design, material/product storage, and other productivity issues.

Pound, Ronald and Smith-Vargo, Linda. "Manufacturing Strategy Can Beat a Bum Rap," *Electronic Packaging & Production,* September 1988, vol. 28(9), pp. 56-61. Illustrates the ideas of manufacturing strategy using two issues: contract assembly and automation.

Priest, J. W. *Engineering Design for Producibility and Reliability.* New York: Marcel Dekker, 1988. Describes various areas of design for producibility. Includes: manufacturing concepts and strategies, manufacturing process analysis, planning, and quality control.

Report of Survey Conducted at Lockheed Missile Systems Division, Sunnyvale CA, Best Manufacturing Practices Program, US Navy, OASN-PI(RD&A), August 1989. Describes Lockheed Missile Systems Division's best practices, problem areas, and industry-wide problems as they relate to critical-path templates of DoD 4245.7-M.

Report of Survey Conducted at Northrop Aircraft Division, Hawthorne, CA, Best Manufacturing Practices Program, US Navy, OASN-PI(RD&A), March 1989. Describes Northrop Aircraft Division's best practices, problem areas, and industry-wide problems as they relate to critical-path templates of DoD 4245.7-M.

Report of Survey Conducted at TRW Military Electronics and Avionics Division, San Diego, CA, Best Manufacturing Practices Program, US Navy, OASN-PI(RD&A), March 1990. Describes TRW MEAD Systems Division's best practices, problem areas, and industry-wide problems as they relate to critical-path templates of DoD 4245.7-M.

Royce, William S. *Is Manufacturing Obsolete?,* Business Intelligence Program, Report No. 83-800, Menlo Park, CA: SRI International, 1983. Emphasizes the need of high-technology and basic industries to work together to prevent the demise of "smokestack America."

Scherkenbach, William W. *The Deming Route to Quality and Productivity: Road Maps and Roadblocks.* Washington, DC: CEEPress Books, 1988. Provides a collection of the author's observations, ideas, and interpretations when implementing the Deming philosophy.

Schmitt, Eric. "Pentagon Scraps $57 Billion Order for Attack Plane," *New York Times,* January 8, 1991, pp. 1,18. Discusses the announcement of the A-12 program cancellation and what it means to defense contractors.

Schonberger, Richard J. *World Class Manufacturing.* NY: Free Press, 1986. Describes the success stories of nearly 100 corporations that have adopted the just-in-time production and total quality control strategies.

Schonberger, Richard J. *World Class Manufacturing Casebook.* NY: Free Press, 1987. Contains case studies of 26 companies' JIT and TQC implementation experiences.

Schuch, Linda K. "TI Does it Right the First Time," *Assembly Engineering,* September 1989, pp. 13-17. Describes a missile assembly facility Texas Instruments developed and the continuous-improvement strategy based on the total quality management principles.

Schwarz, Walter H. "Make It or Buy It?," *Assembly Engineering*, August 1989, pp. 27-29. Describes contract manufacturing and the services provided by contract assemblers. Provides tips on selecting a contract assembler.

Sheridan, John H. "The New Luddites?," *Industry Week*, February 19, 1990, pp. 62-63. Describes the results of a survey of senior-level American managers indicating growing disenchantment with advanced manufacturing technology and the possible causes for the disenchantment.

Sheridan, John. H. "What Makes A Winner?," *Industry Week*, May 21, 1990, pp. 30, 34. Describes Ernst & Young competitiveness study that includes a prescription for marketplace success.

Skinner, Wickham. "A Strategy for Competitive Manufacturing," *Management Review*, No. 76, August 1987, pp. 54-56. Describes nine key areas where U.S. manufacturers are trying to improve and compete in the global market. Describes some important trade-off decisions in manufacturing.

Skinner, Wickham. "Manufacturing—Missing Link in Corporate Strategy," *Harvard Business Review*, May/June 1969, pp. 136-145. Describes the concept of manufacturing strategy and its relationship to other strategies.

Sourwine, D. A. "Improved Product Costing: A Look Beyond Traditional Financial Accounting," *Industrial Engineering*, July 1990, pp. 34-37. Addresses the erroneous decisions caused by several antiquated cost measurement techniques and provides solutions to the problems.

Spurgeon, Susan P., Marcinko, Frank, Mengele, Martin J., Lyman, Richard C. "QPL or QML—A Quality Trilogy Approach," *1990—ASQC Quality Congress Transactions—San Francisco*, 1990, pp. 189-193. Describes a structured approach for achieving manufacturing line certification to MIL-M-38510 in three phases: planning, control, and improvement. Provides MIL-M-38510 Quality Assurance Requirements.

Starr, Barbara, "Foreign Comparative Testing: Looking for the Best," *Jane's Defence Weekly*, vol. 15(4), January 26, 1991, pp. 117-122. Describes the FCT program that assesses foreign equipment to fill gaps and how it can be used to stretch budgets.

Stevenson, Howard H. "Defining Strengths and Weaknesses," *Sloan Management Review*, Spring 1976, p. 65. Subcommittee on Management Quality Change. *Staying Alive: Managing the Process of Change for Quality Improvement*. Dearborn, MI: American Supplier Institute, 1989.

Sutton, George P. *Computer-Assisted Process Planning*, Business Intelligence Program, Report No. 765, Menlo Park, CA: SRI International, 1988. Describes the importance of CAPP to the development process. Discusses current practices, process planning technology, the market for CAPP, and implementation considerations.

Teixeira, John. "Concurrent Engineering," paper presented at the 4th Annual Best Manufacturing Practices Workshop, September 11, 1990, Scottsdale AZ. Addresses the Westinghouse Electric Systems Group efforts in concurrent engineering and discusses the PAC/TAC strategy.

Thompkins, James. A. *Winning Manufacturing*. Norcross, GA: Industrial Engineering and Management Press, 1989. Describes the totality of manufacturing and identifies the critical requirements for "winning manufacturing."

Thompson, A. A., Strickland, A. J. *Strategy and Policy: Concepts and Cases*. Plano, TX: Business Publications, 1981.

Touche Ross. *Operating Principles for the 1990s Phase 1: Assessment of the Elements Comprising World Class Manufacturing*. Ann Arbor, MI: National Center for Manufacturing Sciences, June 89. Describes the infrastructural components of manufacturing strategy and operations.

Usher, John S. et. al. "Redesigning An Existing Layout Presents a Major Challenge—And Produces Dramatic Results," *Industrial Engineering*, vol. 22(6), June 1990, pp. 45-49. Presents a step-by-step method to redesign existing layouts and the challenges redesign poses.

Varzandeh, Dr. Jay and Pickens, Mark. "JIT: Excellence in Manufacturing," *P&IM Review with APICS News*, March 1990, vol. 10(3), pp. 38-40. Discusses misconceptions and mistakes concerning JIT, and presents a case study illustrating many common problems in its implementation.

Vasilash, Gary S. "Manufacturing by Wire," *Production,* June 1990, pp. 85-89. Describes the process through which Northrop Hawthorne production facility became paperless.

Veilleux, Raymond F. and Petro, Louis W. *Tool and Manufacturing Engineers Handbook, Volume 5: Manufacturing Management,* Dearborn, MI: SME, 1988. Provides a comprehensive discussion on manufacturing management as it relates to such topics as strategic planning, cost estimating, management philosophy, CIM, and quality.

"Why the A-12 was Cancelled," *Jane's Defence Weekly,* vol. 15(6), February 9, 1991, p. 175. Presents an account of the A-12 program and some major reasons why the program was cancelled.

Westinghouse. *Westinghouse/AFSC partners in IMIP.* Westinghouse Brochure. Presents Westinghouse's IMIP implementation projects.

Wheelright, Steven C. "Manufacturing Strategy: Defining the Missing Link," *Strategic Management Journal,* vol. 5(1), 1984. pp. 77-91. Provides a framework among corporate, business unit, and functional strategies to achieve manufacturing as a competitive advantage.

Wheelright, Steven C. and Hayes, Robert H. "Competing through Manufacturing," *Harvard Business Review,* January-February 1985, pp. 99-109. Outlines the four stages in manufacturing's strategic role, and the key choices and managerial challenges at each stage.

Williams, D. J. *Manufacturing Systems.* New York: Halstead Press, 1988. Describes manufacturing systems approaches, factory layout, automated factory and systems, and computer control of manufacturing systems.

Wright, Dennis J. "Avionics Manufacturing in the 1990s," *QUEST,* TRW Space & Defense Sector, vol. 13(1), Summer 1990, pp. 2-20. Discusses the integral concepts TRW considers critical to a world-class computer-integrated manufacturing facility, such as the MEAD facility.

Zeimer, D. R. and Maycock, P. D. "A Framework for Strategic Analysis," in *Corporate Strategy and Product Innovation,* Robert R. Rothberg Ed., NY: Free Press, 1976. Summarizes the entire range of activities in determining alternatives for the strategic allocation of resources.

Parts Selection and Defect Control

1

Introduction

To the Reader

This part includes four templates: Parts and Materials Selection, Piece Part Control, Manufacturing Screening, and Defect Control.

The templates, which reflect engineering fundamentals as well as industry and government experience, were first proposed in the early 1980s by a Defense Science Board task force of industry and government leaders, chaired by Willis J. Willoughby, Jr. The task force sought to improve the effectiveness of the transition from development to production. The task force concluded that most program failures were due to a lack of understanding of the engineering and manufacturing disciplines used in the acquisition process. The task force then focused on identifying engineering processes and control methods that minimize technical risks in both government and industry. It defined these critical events in design, test, and production in terms of templates.

The template methodology and documents

A template specifies:

- areas of high technical risk
- fundamental engineering principles and proven procedures to reduce the technical risks

Like a classical mechanical template, these templates identify critical measures and standards. Use of the templates makes it likely that engineering disciplines will be followed.

The task force documented 47 templates and in 1985 the templates were published in the DoD *Transition from Development to Production*

(DoD 4245.7-M) manual.[1] The templates primarily cover design, test, production, management, facilities, and logistics. In 1989, the Department of Defense added a 48th template on Total Quality Management (TQM).

In 1986, the Department of the Navy issued the *Best Practices* (NAVSO P-6071) manual,[2] which illuminated DoD practices that compound problems and increase risks. For each template, this manual describes:

- potential traps and practices that increase the technical risks
- consequences of failing to reduce the technical risks
- an overview of best practices to reduce the technical risks

The intent of the *Best Practices* manual is to help practitioners become aware of the traps and pitfalls so they do not repeat them.

The templates are the foundation for current educational efforts

In 1988, the government initiated an educational program, "Templates: Professionalizing the Acquisition Work Force," with courses and this series of books to increase awareness and improve the implementation of the template concept.

The key to improving the DoD's acquisition process is to recognize that it is an industrial process, not an administrative process. This change in perspective implies a change in the skills and technical knowledge of the acquisition work force in government and industry. Many in this work force do not have an engineering background. Those with an engineering background often do not have broad experience in design, test, or production. The work force must understand basic design, test, and production processes and associated technical risks. The basis for this understanding should be the templates since they highlight the critical areas of technical risk.

The template educational program meets these needs. The program consists of a series of courses and technical books. The books provide background information for the templates. Each book covers one or more closely related templates.

How the parts relate to the templates. Each part describes:

---

[1]Department of Defense. *Transition from Development to Production.* DoD 4245.7-M, September 1985.

[2]Department of the Navy. *Best Practices: How to Avoid Surprises in the World's Most Complicated Technical Process.* NAVSO P-6071, March 1986.

- the templates, within the context of the overall acquisition process
- risks for each included template
- best commercial practices currently used to reduce the risks
- examples of how these best practices are applied

The books do not discuss government regulations, standards, and specifications, because these topics are well-covered in other documents and courses. Instead, the books stress the technical disciplines and processes required for success.

Clustering several templates in one book makes sense when their best practices are closely related. For example, the best practices for the templates in this part interrelate and occur iteratively within design and manufacturing. Designers, suppliers, and manufacturers all have important roles. Other templates, such as Design Reviews, relate to many other templates and thus are best dealt with in other books.

Courses on the templates.   This series of books is designed to be used either in courses or as stand-alone documents. An introduction to the templates and several technical courses are available. The courses use lectures and other proven instructional techniques such as videotapes, case studies, group exercises, and action plans.

The template educational program will help government and industry program managers understand the templates and their underlying engineering disciplines. They should recognize that adherence to engineering discipline is more critical to reducing technical risk than blind obedience to government standards and administrative rules. They should especially recognize when their actions (or inactions) increase technical risks as well as when their actions reduce technical risks.

The templates are a model

The templates defined in DoD 4245.7-M are not the final word on disciplined engineering practices or reducing technical risks. Instead, the templates are a reference and a model that engineers and managers can apply to their industrial processes. Companies should look for high-risk technical areas by examining their past projects, by consulting their experienced engineers, and by considering industry-wide issues. The result of these efforts should be a list of areas of technical risk which becomes the company's own version of the DoD 4245.7-M and NAVSO P-6071 documents. Companies should tailor the best practices and engineering principles described in the books to suit their particular needs. Several military suppliers have already produced manuals tailored to their processes.

*Materials are the heart of engineering
design...Selecting the optimum materials
and the subsequent processes necessary to
turn them into finished parts calls for
considerable skill on the part of the
designer, a process made more complex by
the continuous introduction of new or
improved materials.*[3]

Parts and materials are selected on the basis of function, performance, reliability, quality, maintainability, methods of manufacture, and cost. Function is the most fundamental reason but the others are also important. Cost refers not only to the cost of the materials but also to the "cost of ownership"—the cost of inventory, special assembly equipment, rework, and delays.

The U.S. Department of Commerce's 1984 survey data on cost distribution show the importance of parts selection and defect control. With commercial systems, the cost of parts and material is about 60% of the cost of the manufactured system. Labor is typically about 20% to 30%.

Figure 1 shows where to find more and more details about risks, best

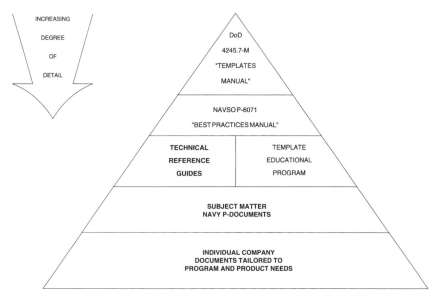

Figure 1 Resources on the acquisition-process templates.

---

[3]Ball, Graham. "How Well Do Materials Meet Designers' Real Needs?" *Materials: Proceedings of Materials Selection and Design.* London, England, July 1985, p. 208.

practices, and engineering principles. Participants in the acquisition process should have copies of these documents.

Figure 2 shows that the percentages vary from industry to industry.[4] The cost of parts and materials is about 70% of the manufacturing cost in the motor vehicle industry, but about 45% in the shipbuilding industry. With military systems, the percentages may range from about 10% to about 30%.

Military systems are less sensitive to parts costs, but the total cost of the military systems tends to be higher than commercial systems. Thus, parts selection and defect control are vital in military systems as well as in commercial systems. With both commercial and military systems, false economy at the parts level as well as parts proliferation can cause horrendous costs later in the system's life cycle.

The four templates discussed in this part identify critical areas of risk. They emphasize the importance of carefully selecting parts and materials and controlling defects in piece parts and systems. Screening and prevention are key methods of control. The following pages define each template and give an overview of the risks and their consequences.

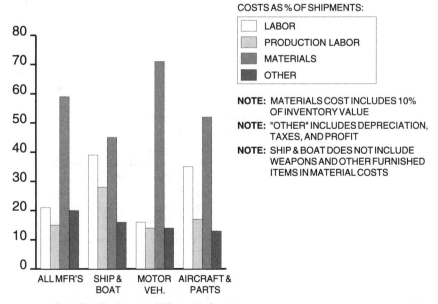

Figure 2 Cost distributions in different industries.

***

[4]Nevins, James L. and Whitney, Daniel E., Eds. *Concurrent Design of Products and Processes.* New York: McGraw-Hill, 1989, pp. 39–40.

The Procedures chapter of this part describes engineering fundamentals and best practices to minimize these risks. The Application chapter gives examples of how to use the best practices that reduce these technical risks.

Parts and Materials Selection Template

Designers select parts and materials to meet specified requirements for functionality, performance, reliability, quality, producibility, cost, etc. Designers need to become familiar with the current best practices to avoid poor selections. Table 1 gives risks in parts and materials selection. The Procedures chapter of this part describes best practices to avoid those risks.

Risk: No preferred-parts list

Preferred parts are parts whose quality and reliability are well-known and which the company may already be using for other systems. Without a preferred-parts list, designers choose parts unsystematically, even though the preferred parts would perform as well or better. The result is a proliferation of nonstandard parts, varying in performance and reliability.

Nonstandard parts may have wide-ranging consequences for manufacturing, purchasing, and logistics. Manufacturing engineers must cope with parts that require a variety of assembly methods and tooling. Inventory costs may increase. Ease of automation may decrease. Pur-

TABLE 1   Risks and Consequences in Parts and Materials Selection

Risks	Consequences
No preferred-parts list at start of development	A proliferation of nonstandard parts, varying in quality and reliability
Obsolete parts selected	Without a preferred-parts list, designers may choose parts that are obsolete or hard to obtain
New technology parts selected	Without a history of proven reliability and mature technology, parts' reliability may be questionable
Parts unsuitable for particular applications	System fails because design can not withstand stresses during use, transportation, or storage
Incomplete or inaccurate thermal analysis data on part operating temperature and vibration	System reliability is lowered because parts are not selected and optimally placed using thermal and mechanical analysis data

chasing representatives must deal with many different suppliers, making it hard to monitor timely delivery and quality and obtain volume discounts. Logistics specialists must provide spares for many different parts and find storage space for them.

When a preferred-parts list is available at the start of full-scale development, designers can select preferred parts or obtain engineering justification for any nonapproved part. (The justification process should be efficient to avoid unnecessary delays.) Designers can follow design-for-assembly practices in reducing the number of parts and choosing standardized parts that can be used in many assemblies and subassemblies.

Risk: Obsolete, hard-to-obtain, or new technology parts

Without a preferred-parts list, designers may choose obsolete or sole-sourced parts that may be difficult to obtain. Preferred-parts databases help designers identify obsolete parts and how long parts will be available.

Without a preferred-parts list, designers may choose the latest part or an exotic part rather than one with proven reliability and mature technology. Parts should undergo an independent assessment before being put on the preferred-parts list. Information sheets from suppliers may not give complete information.

Risk: Parts unsuitable for particular applications

In selecting parts, designers often fail to consider how the part will be used—what stresses the part will encounter in use and during storage and transportation. Often the stresses during storage and transportation are greater than during use. Likely stresses include shock, vibration, and changes in temperature.

Designers also fail to make their designs able to tolerate wide environmental variations. Similarly, they fail to use *derating* techniques to choose parts manufactured to withstand more than the likely operational, storage, and transportation stresses. With larger and stronger parts, there is a safety margin because parts will be stressed only to a percentage of their capability. Parts can then better withstand hidden stresses in manufacturing, testing, operation, storage, and transportation.

To minimize these risks, companies should establish for all engineers a design policy covering parts and material selection. This policy should set specific limits on allowable stresses.

Risk: Incomplete or inaccurate analyses

Designers' thermal and mechanical analyses are often incomplete and may be based on subjective estimates and probabilities. Without simulations and thermal surveys to measure part operating temperatures, parts may be located too far from heat sinks, thereby lowering the reliability of the system. Mechanical deflection and vibration-related effects are often not simulated during the design process. It is much better to incorporate the results of thermal, mechanical and stress analyses into the design early. If design deficiencies are not found until later during testing, design changes are more difficult and costly. Finite-element analysis and finite-difference analysis are techniques widely used to analyze stresses and temperature profiles. These techniques produce accurate results.[5] *Design to Reduce Technical Risk* has more details on design analyses.

Piece Part Control Template

Piece part control programs are needed to ensure timely delivery of high-quality, high-reliability parts. Supplier certification programs with timely feedback and exchange of information are a way to ensure incoming parts meet specifications. Table 2 gives risks in piece part control. The Procedures chapter gives best practices to avoid the risks.

Risk: Lack of a formal piece part control program during development

Without a formal parts control program, requirements may not flow down to subcontractors and suppliers. Parts and subassemblies may not meet specifications for quality, reliability, and timeliness. Incoming parts may not meet specifications and thus cause scrap, rework, and higher inventory costs as manufacturing tries to work around the defective parts.

Piece part control depends on parts and material selection. Unless designers have access to a preferred-parts list and a formal parts control program, they are likely to select from suppliers whose parts' quality and reliability are not optimal. They may unknowingly choose suppliers with questionable financial stability. They may unknowingly choose suppliers who have a high reject rate or who fail to meet delivery deadlines. They may choose distributors' parts that are risky because their origin may be uncertain and defects may be introduced during storage and transportation.

[5]*Reliability Assessment Using Finite Element Techniques.* RADC Technical Report, TR-89-281, 1989.

TABLE 2   Risks and Consequences in Piece Part Control

Risks	Consequences
Lack of formal piece part control program during development	Requirements may not flow down to subcontractors and suppliers, and thus parts may fail to meet specifications, causing scrap and rework
Failure to qualify and certify suppliers	Parts may vary in quality and reliability, especially if they come from distributors
Poor source management	Piece part quality and reliability may be low if source management techniques are not used (e.g., qualify and certify suppliers, do source inspection, and do receiving inspection if necessary)
Blind dependence on preconceived standards	Even MIL-STD parts may have high defect rates if they are not manufactured with stable process controls
Poor handling during screening or rescreening	Poor handling may damage parts due to mechanical damage, electrostatic discharge, or electrical overstress
Inadequate attention to older-technology parts	Older-technology parts (e.g., relays, transformers, inductors, switches, and connectors) may contribute to failure in many systems

Risk: Poor source management

Piece part quality and reliability may be low if companies fail to use the best approach to eliminate marginal devices. Ideally, companies select parts from qualified and certified suppliers. If this is not possible, companies may send their own inspectors to the supplier's site. This source inspection may fail to ensure high quality and reliability especially if there are strained relations between the supplier and the assembly manufacturer or if source inspectors are assigned to many locations and thus can visit sites only infrequently. When done properly, source inspection may be preferable to rescreening at the assembly manufacturer's site because it saves equipment, labor, and the expense and delays due to return of defective parts.

Risk: Blind dependence on preconceived standards

In the 1980s, electronics parts varied widely in electrical and mechanical quality, even those that were manufactured and screened according to military standards such as MIL-M-38510, General Specification for

Microcircuits. To increase system reliability, some assembly manufacturers began to rescreen incoming parts, especially integrated circuits. Screening to find parts with actual and hidden defects requires resources of time and money. Poorly applied, these resources fail to add value. Screening is ineffective when it does not remove marginal devices. Screens should not be chosen arbitrarily without regard to how they will be used. Screens should be tailored to bring out actual and hidden defects. The results of screening should be used to find root causes and corrective actions to prevent the defects.

Risk: Rescreening

Rescreening should be a last resort. Rescreening may reduce a device's reliability by inducing additional failures. Poor handling may damage parts due to mechanical damage, electrostatic discharge, or electrical overstress. Rescreening adds no value when the supplier's process is stable and producing near-zero defects. Screening at the part level should be done during original manufacture as part of a process control program. Rescreening should be performed only on a temporary basis. An example is using nonqualified vendors until qualified vendors can be found.

The resources spent on rescreening to ensure usable parts could be better spent:

- monitoring the supplier's process control, testing, and piece part screening

- maintaining records on defect rates from various suppliers

- providing feedback to suppliers and designers on yield and defect rates

- setting up alliances with certified suppliers who have a record of timely delivery of good parts

Selecting suppliers is critical. Companies must decide what data they need to select suppliers who have stable processes and near-zero defects. The Procedures chapter describes what data Texas Instruments is using. The Application chapter describes the data and procedures AT&T and Magnavox are using. All the companies use data to make decisions, although their procedures vary.

Manufacturing Screening Template

Manufacturing screening stresses assemblies such as boards, units, and subsystems to stimulate assembly, process, and installation defects. Part defects should be detected before they are assembled into boards or units. If part defects are found during manufacturing screening, the part manufacturing and part screening processes should be investigated.

Manufacturing screening is often called environmental stress screening. Temperature cycling and random vibration are the most common screens. The screening is tailored to provide appropriate limits, number of cycles, and rate of changes for temperature cycling, vibration, and shock. Table 3 gives risks that arise from a lack of understanding of environmental stress screening. The Procedures chapter describes best practices to avoid the risks.

Risk: Failure to understand environmental stress screening

To obtain the greatest benefit with least risk and lowest cost, companies must understand how to use screens to stimulate hidden defects. Without this understanding, screens may not be strong enough to stimulate likely flaws. Or screens may be too strong and unnecessarily damage good units. Overly intense screens induce additional defects and consume too much of a unit's operating life.

TABLE 3   Risks and Consequences in Manufacturing Screening

Risks	Consequences
Failure to understand environmental stress screening	Screens may not be strong enough to stimulate likely flaws, or screens may be too strong and unnecessarily damage good units
Screens not tailored to particular process and technology	Workmanship and quality problems may not be detected if screens are chosen inappropriately, and temperature cycling and random vibration screens are not used to find the most defects at the lowest life-cycle costs
Screening is done at inappropriate assembly levels	Screening done at lower assembly levels may not find latent defects in interfaces or interactions, screening done at higher assembly levels may not prevent expensive rework
Lack of corrective action	Assembly and workmanship defects persist even with stress screening if root cause analysis is not used to prevent future defects

Risk: Screens not tailored to process and technology

Ineffective screens may not contain the appropriate level of intensity, duration, range, rate of change, repetitions, etc. Ineffective screening programs may include temperature cycling but not random vibration. Many workmanship and quality problems may escape detection. Temperature cycling and random vibration screens should be tailored to the particular process and technology to find the most defects at the lowest life-cycle costs.

Screening results should be constantly reviewed. The results should be used to increase, decrease, or modify the screens to make them more effective in reducing defects and life-cycle costs. (Life-cycle costs include the cost of repair in the field.)

Risk: Screening is done at an inappropriate assembly level

It is important to consider the advantages and disadvantages of screening at different levels to decide which level is most cost-effective. Ineffective stress screening fails to prevent expensive rework at higher assembly levels. In general, detection efficiency is best at the highest assembly level (the system level), but the cost per failure is least at the lowest assembly level (i.e., the circuit card level). It may be more effective to use stress screening at two or more levels to reduce expensive rework while maintaining detection efficiency.

Risk: Lack of corrective action

Even with effective screening programs, assembly and workmanship defects may persist if root causes and corrective actions are not found. The results of root cause analyses may suggest improvements in the process control systems. These improvements should then be verified.

Defect Control Template

Assembly and workmanship defects result in higher production costs, increased process times, and expensive rework. The best way to prevent these defects is with a disciplined defect control program. Table 4 gives risks in defect control. The Procedures chapter describes best practices to avoid the risks.

Risk: Failure to commit to prevention

Companies should commit to prevention as the prime ingredient of a sound defect control program. Ineffective programs often overlook key

TABLE 4   Risks and Consequences in Defect Control

Risks	Consequences
Failure to commit to prevention	Defects will persist if the emphasis is on short-term fire-fighting rather than on prevention of future defects (e.g., identify critical procedures, equipment, personnel, training)
Poor use of yield and defect data	Defect reports may not be used because they are lengthy and out-of-date when issued
	Factory engineers put too much emphasis on the statistical implications of defect and yield data
	Data and charts are accessible only to factory engineers rather than being visible and understandable by all workers
Lack of root cause and failure analysis programs	Defects continue to occur at a high rate
Lack of timely repair	Delays in identifying, analyzing, and correcting problems may make the results no longer meaningful

elements in preventing and controlling defects—designs, procedures, equipment, personnel, and training. Thus, defects continue to cause delays and disruption. Poor designs may persist when companies fail to make designs less susceptible to variations in materials, processes, testing, or environmental conditions.

Risk: Poor use of yield and defect data

Defect reports are often lengthy and out-of-date when issued. Delays and errors may occur because manual entry of data is cumbersome and time-consuming. Many programs could improve the failure reporting system with an automated tracking system for more timely information. For example, bar codes and scanners can help track and monitor yields and defects.

Factory engineers often overemphasize the statistical implications of defect and yield data instead of using data to trigger corrective actions to improve out-of-control processes. Data and charts are accessible only to factory engineers. Charts on current yields, defect rates, and improvements are not displayed to make them visible and understandable by all workers. Shop-floor workers are not trained to interpret the data.

Successful companies are now training all employees to use statistical process control techniques. These techniques help them collect and use data for continual improvement.

Risk: Lack of root cause and failure analysis

Defects often continue to occur at a high rate if there is too little corrective action to find the root causes of defects and then solve and prevent them. Often, too little feedback goes to the designers and suppliers.

Failure analysis is often "too little, too late." Teams do not include representatives from design, testing, manufacturing, and other areas with key information. Teams are not given the time and resources to investigate failures. They neglect to reproduce the failure. They do not test to find the root cause, especially if the failure occurs intermittently. They also fail to identify and verify all the causes of failure.

The data on the defect's source or causes should be used to correct the process or system to prevent defects.

Risk: Lack of timely repair

Because of time pressures in manufacturing environments, faulty units or assemblies are often set aside for later testing and repair. Often people give higher priority to production testing than troubleshooting defective units. This often results in delays in identifying, analyzing, and correcting problems. Priorities must be constantly reviewed. These reviews ensure the proper balance between production and troubleshooting and help people make trade-offs between the cost of fixing the process vs. the costs of increased overtime, schedule delays, and field failures.

Rationale for an Integrated Parts Control Program

To minimize risks, an integrated parts control program is needed. This program should be multidisciplinary, involving designers, manufacturing engineers, suppliers, purchasing representatives, and subcontractors. Improvements in parts selection and defect control will result in fewer rejects, less rework, less downtime, and fewer failures in the field. These benefits far outweigh the savings in choosing a low-cost bidder with marginal quality.[6] For most large projects, the parts control program can be managed most effectively by a computerized database system.

---

[6]Maass, Richard. *World Class Quality: An Innovative Prescription for Survival.* Milwaukee, WI: ASQC Quality Press, 1988, p. 13.

Below are best practices for design, manufacturing, suppliers, purchasing, and subcontractors to achieve an integrated parts control program.

Best practices for design

- Analyze functionality, reliability, maintainability, testability, and cost requirements.
- Use preferred-parts lists and consult databases and libraries to find standard designs for reuse.
- Make sure part capability matches design needs.
- Analyze failures to find root causes and continuously improve design of products and processes.

Best practices for manufacturing

- Participate in design decisions early.
- Ensure designs reflect producibility and testability requirements.
- Review electrical-test specifications for adequacy.
- Set up a formal piece part control program.
- Decide what stress screening is appropriate and continually assess data to see if screens can be eliminated.
- Establish and maintain an effective failure review and corrective action system.
- Set up supplier and manufacturer quality-improvement teams.
- Give feedback to designers and suppliers.
- Select manufacturing equipment, test equipment, special processes, and resources for high quality and high yields with continual process improvement.

Best practices for suppliers

- Ensure requirements are understood.
- Participate in partnership agreements.
- Use statistical process control and feedback for continual improvements.
- Establish and nurture quality-improvement teams.
- Use failure analysis to improve designs, processes, and part performance.

Best practices for purchasing

- Procure parts from qualified suppliers when possible.
- Develop partnership agreements.
- Help supplier continuously improve quality and cost.

Best practices for subcontractors

- Ensure requirements flow down.
- Participate in design reviews with prime contractor.
- Set up long-term contracts for mutual benefit.

2

Procedures

The four templates in this part have been integrated into six steps as Figure 3 illustrates.

Each template is discussed in Step 1: Create an Integrated Parts Management Strategy. Parts and Material Selection is discussed in

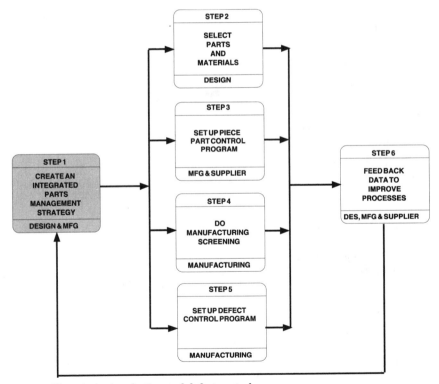

Figure 3 Steps in parts selection and defect control.

Step 2: Select Parts and Materials. Piece Part Control is discussed in Step 3: Set Up a Piece Part Control Program. Manufacturing Screening is discussed in Step 4: Do Manufacturing Screening. Defect Control is discussed in Step 5: Set Up a Defect Control Program. Each of the four templates is also discussed in Step 6: Feed Back Data to Improve Processes.

Figure 3 also shows at the bottom of each box who is primarily responsible for carrying out each step: design, manufacturing, or suppliers.

Step 1: Create an Integrated Parts Management Strategy

To be effective, a parts management strategy should be wide-reaching and truly integrated with designers, manufacturers, quality and reliability engineers, suppliers, subcontractors, and purchasing representatives all working together to meet customer needs. An integrated strategy gives continuity throughout the product life cycle. Systems engineers, designers, and other early decision makers are more likely to consider the needs of later decision makers—testing, manufacturing, logistics, and maintenance. This strategy works only with well-planned communication and feedback loops.

Create integrated databases

Designers, manufacturers, subcontractors, project management, purchasing representatives, and supplier partners work more smoothly if they all have access to database information. With all organizations working together, they can use the databases to achieve mutually agreed-upon goals. Thus, suppliers agree not to use the information for individual or competitive goals. Information must be updated continually to ensure that it remains useful and everyone works with current information.

Benefits. The benefits of integrated databases include:

- better decision making when all participants have access to up-to-date information
- quicker corrective action
- higher productivity from faster development, less rework, and lower maintenance costs

Use a series of linked databases. A key element of an integrated parts management strategy is whether information is accessible to each participant. Rather than one large parts database, the most manageable databases are linked databases that have different sets of information for a variety of purposes. For example, one database may include qualification data, another has flammability data, and another has packaging information. Yet all these databases can share data and link with each other.

A few critical index parameters can be used to link these databases to each other. These parameters may include:

- specification number (generic code number rather than specific manufacturer's number)
- supplier by location (locations may have different processes and different quality)
- part description or family (e.g., ceramic capacitor)

Examples of linked databases.   Below are examples of information in specific databases.

A design database should include:

- preferred parts (based on performance, producibility, reliability, power dissipation, availability, cost)
- component selection tools and libraries compatible with design processes and tools
- reliability data (e.g., component failure rates, procedures to calculate failure rates, data from manufacturers' data books)

A supplier database should include:

- supplier's approval rating based on their quality, reliability, and timeliness
- supplier's quality and reliability data
- schedule of audits

A manufacturing database should include:

- forecasts
- producibility and design-for-manufacture requirements
- yields and defect rates for parts, boards, and units
- results of failure analyses of parts that had to be removed from circuits and apparatus

The failure analysis indicates why the failure occurred—defective parts, design-rule violations, missed specifications, wrong applications, inadequate design margins, accidental abuse, or incorrect assembly. These problems are then traced back to their source.

Ideally, feedback is then given to the designer or the supplier. With feedback on troublesome parts, designers understand the specific causes of failure and can avoid them perhaps by adhering more closely to design rules, specification changes, or requalifications. With information on specific defects traceable to the location on the board, designers may be able to avoid these problems in their future designs.

A life-cycle database should include:

- parts that are discontinued but will be needed to support a product over its life cycle

- information collected during the product's life cycle

Feeding all these databases are the quality and reliability databases, which contain information on a supplier's quality and the defects later encountered.

These databases strengthen the interrelationships among design, reliability, quality, purchasing, suppliers, manufacturing, and accounting. Many companies use integrated databases, including AT&T, Texas Instruments, General Dynamics, and GTE. The Application chapter describes how integrated databases are used at McDonnell Douglas Helicopter. For more information on databases and computer-aided tools, see *Design to Reduce Technical Risk*.

Step 2: Select Parts and Materials

Establish preferred-parts lists

Successful companies have lists of preferred parts and materials that have met criteria for performance, reliability, timely delivery, and reasonable cost. Preferred-parts lists should be continually updated with historical data on delivery schedules, quality, reliability, performance, and costs. Companies that buy large quantities of standardized parts and materials see improvements in inventory costs, quality, ease of automation, and materials handling. Tooling and assembly costs are also less.[7]

---

[7]Priest, John. *Engineering Design for Producibility and Reliability*. New York: Marcel Dekker, 1988, p. 181.

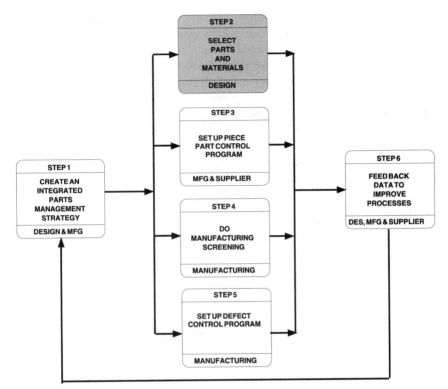

Figure 4 Steps in parts selection and defect control.

In using preferred parts and materials lists, designers are better able to follow good design rules, including design-for-assembly practices. These practices include reducing the number of parts and choosing standardized parts that can be used in many subassemblies and products.[8]

Before developing a new design, designers look in an integrated design database for a part that can be reused or modified to meet new design requirements. To make this possible, a database of coded designed parts must be created. Each part should have its own part number tied to its drawing and its own code number tied to a family of parts with similar attributes.

Designers can improve producibility by selecting parts from a few distinct part families with common design and manufacturing characteris-

[8]Nevins, James L. and Whitney, Daniel E., Eds. *Concurrent Design of Products and Process.* New York: McGraw-Hill, 1989, p. 205.

tics. In *group technology,* similar parts are identified and manufactured together in a common production-line environment. Shorter intervals and improved quality usually result from group technology.[9]

In general, designers select parts from two broad categories. In the first category are the primary parts that provide basic capabilities. In the second category are support hardware and parts used to assemble the item or to provide secondary functions. Examples of support hardware and parts include brackets, hoses, nameplates, passive electrical parts, and indicator meters. These secondary parts offer the most potential for design standardization and improvements in efficiency from the use of group technology.[10]

Submit stocklists for approval

In some companies, designers submit stocklists (lists of the parts their design needs) to component managers who check to see if the parts are in the preferred-parts database. In many companies, component managers provide data to help the designers select parts. The data include power dissipation, reliability, and cost. Component managers get reliability data from factory and corporate databases and manufacturers' data books. They carefully evaluate these data.

Component managers, who may be reliability engineers, occasionally conduct reviews to audit the preferred parts for reliability, availability, manufacturability, and cost. Recommended changes must go through a formal change approval process. Often, the component manager has the responsibility to update the preferred-parts lists.

Select from component equivalents. Component managers usually work with preliminary stocklists, which are mostly generic codes. A single generic code can correspond to, or be functionally equivalent to, several manufacturers' devices: e.g., parts made by Motorola, Intel, and Texas Instruments may have the same functions. Equivalent device codes have the same generic code but the packaging, testing, and qualification level may vary. Circuit designers use generic device codes to avoid unnecessary restrictions at the early stages of design.

Component managers often approve the designers' selection of parts from among equivalent parts. Circuit designers may select specific device codes for some devices (memory chips and gate arrays) but they typically specify generic codes for most integrated circuits. Circuit

---

[9]Priest, John. *Engineering Design for Producibility and Reliability.* New York: Marcel Dekker, 1988, p. 180.

[10]Snead, Charles S. *Group Technology: Foundation for Competitive Manufacturing.* New York: Van Nostrand Reinhold, 1989, p. 121.

designers usually specify preference among equivalent parts based upon power, reliability, availability, quality, supplier responsiveness, size, weight, cost, etc.

On military contracts, designers often have narrower choices. They may be obligated by contract to use specific parts. For example, on many contracts designers must choose Joint Army Navy (JAN) parts that meet precise specifications. The recent defense initiatives have encouraged efforts to improve and streamline the component selection process. An ad hoc committee with representatives from the semiconductor industry and the defense community is developing an applications guide for contract officers that gives guidelines and trade-offs for using parts in various applications.

Consider quality and reliability

The terms "quality" and "reliability" are often used interchangeably as though they were the same attribute. They are different, however. Quality is the composite of all required attributes, including performance. It must be built in by means of stable design and manufacturing processes. Quality is the number of good parts or products that arrive at the next user.

Reliability is "quality on a time scale."[11] That is, reliability is the ability "to perform a required function under stated conditions for a stated period of time."[12] To produce a reliable device, the designer and manufacturer consider the interactions of the design, how well it is executed during manufacture, and the environmental stresses under which the product will function.

Reliability data should include information on failure rates and stress factors. A failure reporting and corrective action system (FRACAS) should include the failure mode, the failure cause, the corrective action, and the effectiveness of the corrective action. The designer can then use these failure rates to predict a device's reliability when it operates under similar conditions. Simulations are useful in predicting reliability. *Testing to Verify Design and Manufacturing Readiness* has more details on reliability.

Consult MIL-HDBK-217 for component failure rates. MIL-HDBK-217 was developed to provide a consistent and uniform database for making reliability predictions when there is no reliability experience for a system.

---

[11] Hnatek, Eugene R. *Integrated Circuit Quality and Reliability*. New York: Marcel Dekker, 1987, p. vi.

[12] Klinger, David J., Nakada, Yoshinao, and Menendez, Maria, Eds. *AT&T Reliability Manual*. New York: Van Nostrand Reinhold, 1990, p. 2.

MIL-HDBK-217 gives failure rates for different part families and discusses two procedures for determining reliability: the parts count method and the parts stress method. The parts count method is used early in the conceptual and design phases. In general, the parts count method uses the parts failure rates to estimate the reliability of an assembly. The parts stress method provides input to design trade-off decisions. It gives different formulas for predicting the reliability of microelectronic devices, discrete semiconductors, and other types of components. For most devices, the failure rate is calculated from the circuit complexity, the package complexity, temperature, quality, etc.

Cautions in using MIL-HDBK-217. This military handbook is sometimes cited as the industry standard for component failure rates because there is no other universally available source of data. The government may require contractors to use models to evaluate the reliability of electronic equipment for government use.

The user should be cautious, however, in using MIL-HDBK-217. Designers can use the failure rates to compare alternative designs but not to predict the field reliability of a device or a system. Field failures, for example, are more often caused by assembly or workmanship failures (e.g., connectors and solder joints) than by electronic-component failures. MIL-HDBK-217 does not address reliability-related problems induced by operators, workmanship, software, or maintenance. The handbook is not meant to predict field reliability and does not do it very well.[13]

Another problem with MIL-HDBK-217 is that the life of solid-state devices and other electronic devices can only be approximated by an exponential distribution. The exponential distribution is a good starting point to estimate times between failures for a system. But additional surveys and analyses are needed to verify the final design.

As the complexity of microcircuit devices increases, the MIL-HDBK-217 models and estimates will become less useful for designers. If possible, designers should substitute known field failure rates of devices into simulations and surveys to prove the design is capable of meeting reliability requirements.

Use other sources of reliability data. Reliability data are useful for selecting components, for budgeting, appraising, and improving reliability, and for planning environmental stress screening (which will be discussed in Step 4). Many projects calculate reliability with

---

[13]Morris, Seymour F. *MIL-HDBK-217, Use and Application*. Rome Air Development Center Technical Brief, April 1990.

mathematical models or by actual tests. Each has advantages and disadvantages. Simulations provide results quickly and at relatively low cost, but the validity and accuracy of their results depend on the assumptions put into the model. Actual testing would give more valid results, but it is time consuming, expensive, and often virtually impossible.

Consider producibility

Designers make subtle choices about materials, components, fasteners, coatings, and adhesives, often without considering how their choices will affect manufacturing. If manufacturing engineers participate in these decisions, they can help designers consider producibility as one of the key criteria.[14]

To consider producibility in selecting parts, designers often work with component and manufacturing engineers to:

- minimize the number of different parts with the same function

- select parts that are functional, reliable, and conform to the factory's capabilities (e.g., parts that have low defect rates, can be automatically assembled, can be wave-soldered and cleaned)

- ensure enough parts are available at the right time

These considerations help reduce the total manufacturing cost by decreasing assembly errors, tooling and repair costs, labor time, and part shortages.

Designers can also increase producibility by giving preliminary stocklists to the manufacturing engineers early. With early stocklists, manufacturing engineers and purchasing agents can investigate suppliers' backgrounds and capabilities. They can make sure suppliers test components for high reliability and quality. Purchasing agents can negotiate contracts to make sure components are available "just in time" for assembly.

To ensure producibility and reduce component cost, many companies set up component-control committees and preferred-component databases tailored to specific manufacturing needs. The committee evaluates requests to use nonpreferred parts and decides when exceptions are justified. The committee also audits stocklists to check compliance with preferred-parts lists and producibility and reliability requirements.

---

[14]Nevins, James L. and Whitney, Daniel E. Eds. *Concurrent Design of Products and Processes.* New York: McGraw-Hill, 1989, p. 200.

Consider derating

In derating, a designer ensures that the actual stresses will be less than a percentage of the maximum stress according to the manufacturer's rating. In selecting parts and material, designers consider how the part will be used to make sure the applied stress is less than the manufacturer's rating. For example, the stresses for space applications are sometimes greater than ground applications.

Derating techniques increase the circuit's capacity to resist stress or decrease stress variations. Examples include using fans, heatsinks, or packaging. As a rule of thumb, reliability doubles with each 10-degree decrease in junction temperature. If reducing stress is not feasible, designers select larger or stronger parts that can withstand more stress. These parts provide a safety margin because they will be stressed electrically, mechanically, and thermally only to a percentage of their capability.

Use derating curves. In selecting parts that can withstand the applied stresses, designers use derating curves to find the appropriate derating percentages. At a point in the operating range there is an inflection point where a slight change in stress causes a large increase in the part's failure rate. The appropriate derating percentage should be well below this inflection point.

Figure 5 illustrates a derating curve.[15] On the vertical axis is the derating percentage, which is the actual power rating in watts divided by the manufacturer's rating. On the horizontal axis, the point of inflection is at T_S. After this point, as the ambient temperature (T_A) or case temperature (T_C) increases to the maximum temperature (T_{max}), the derating percentage decreases. The area below the maximum-use rating curve

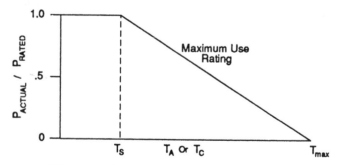

Figure 5 Illustrative derating graph.

[15]Anderson, R. T. *Reliability Design Handbook*. Chicago: ITT Research Institute, 1976, pp. 135–143.

provides the optimum margin of safety with no degradation in reliability expected. In the area above the maximum-use rating curve, however, parts would be overstressed.

Derating curves are available for different part types. These curves are sensitive to changes in temperature, electrical transients, vibration, shock, altitude, and acceleration. For example, the junction temperatures of semiconductors and integrated circuits should not exceed 110°C regardless of the nominal power rating. On a transistor, the collector-to-base voltages should be less than 70% of the manufacturer's value.

Figure 6 shows how the acceptable derating ranges for glass capacitors depend on the application. In general, ground applications may be less stressful than airborne or space applications.[16]

Match the derating technique to the part type and application. The specific derating techniques vary with different types of parts and different applications. For example, capacitors are derated by keeping the applied voltage at a lower value than the voltage for which the capacitor is rated. Semiconductors are derated by keeping the power dissipation below the rated level.

With semiconductors, in addition to reducing part failure, derating reduces the internal operating temperature, thus decreasing the rate of chemical time-temperature reaction that causes part aging. Designers also use functional derating to help them make conservative design decisions with integrated circuits. One decision, for example, is the amount of fanout in a logic circuit, which refers to the number of paths fanning out from one output to various inputs on other devices. If the maximum fanout is to eight input devices, for example, conservative functional derating would restrict the fanout to six input devices. This conservative derating allows the circuits to function even with wide temperature variations and makes the circuits easier to test.

Judge the appropriate derating level. Projects should collect data to derive derating criteria that meet their needs. Too-stringent criteria unnecessarily boost costs; too-liberal criteria increase failure rates.[17] The Rome Air Development Center has developed a derating "slide ruler" that provides derating guidelines for different part types and design requirements.[18]

---

[16]Anderson, R. T. *Reliability Design Handbook*. Chicago: ITT Research Institute, 1976, p. 166.

[17]Priest, John. *Engineering Design for Producibility and Reliability*. New York: Marcel Dekker, 1988, p. 186.

[18]RADC. *Parts Derating Guideline*. AFSC Pamphlet 800-27, Rome Air Development Center, 1983.

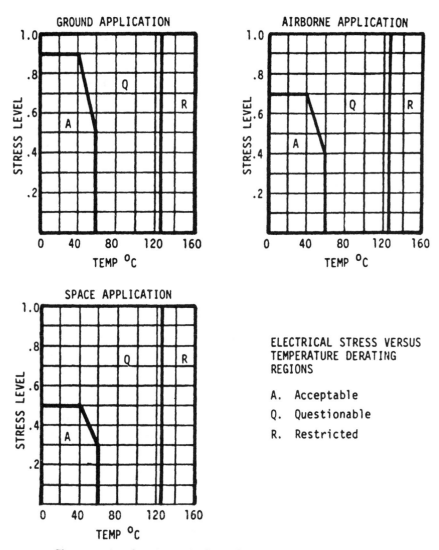

Figure 6 Glass-capacitor derating varies by application.

The advantages of derating often come with some disadvantages: parts having greater capability are often more costly, bigger, and heavier. The designer must decide which trade-offs are most important.[19] The Application chapter gives an example of derating used in the F/A-18 Hornet.

---

[19]Naval Sea Systems Command. *Parts Application and Reliability Information Manual for Navy Electronic Equipment.* TE000-AB-GTP-010, September 1985, p. F-2.

Derating is often used to include safety margins throughout the design as well as the ability to perform additional capabilities in the future. Examples include a power supply that can supply more current than needed, parts that can withstand higher stresses than are applied, a receiver microprocessor that operates at 25% of its rated capacity, and transmitter microcircuits that operate at one-third of their rated voltage.

Using derating principles to add surplus capability or margin to designs contributes more to reliable operation than adding quality improvements to an already high-quality product. The designer judges the appropriate derating level, balancing the need for mission success, cost, schedule, and future enhancements.[20]

Do stress analysis

As stress increases, a component's failure rate increases. Even when a component does not fail, the circuit may fail due to stress-induced reduction in operating margin. One way to prevent these failures is the use of derating criteria. Stress analysis ensures that the derating criteria are met and verifies that the equipment can perform under worst-case conditions.

Stress analysis is difficult with:

- a mix of technologies (e.g., bipolar, CMOS, TTL), which makes interface stress problems more likely and more difficult to detect

- large-scale and very large-scale integrated circuits that make tracing of loads more difficult[21]

Stress analysis is usually done hierarchically. Initially, the stresses on the smallest possible components are analyzed. Then the stresses on higher assembly levels are analyzed. The stress analyses may include mechanical stress (from bending, shear, or torsion), thermal stress (from temperature changes and extreme temperatures), and electrical stress (from power, current, or voltage).

Do thermal analysis

Designers analyze thermal profiles to measure part operating temperatures and compare measured temperatures to derating criteria. Thermal

---

[20]Meinen, Carl. "Reliable Remote Monitoring and Control of Electrical Distribution." *Proceedings of the Annual Reliability and Maintainability Symposium,* 1980, pp. 448–452.

[21]Bannan, M. W. and Banghart, J. M. "Computer-Aided Stress Analysis of Digital Circuits." *Proceedings of the Annual Reliability and Maintainability Symposium,* 1985, pp. 217–223.

analyses are computer simulations that help the designer select parts and materials to avoid thermal coefficient-of-expansion mismatches and other design problems. Thermal analysis is used to select materials for particular applications and to check the quality of incoming raw materials and outgoing products. To make thermal-analysis systems more affordable and flexible, suppliers are using personal computers to analyze and store data.[22] Finite-element analysis is a useful method for analyzing stresses and heat transfer. The results of thermal analyses should be verified with thermal surveys. For more details on stress analysis and thermal analysis, refer to *Design to Reduce Technical Risk*.

Thermal mismatches.   One example of the use of thermal analysis is with surface mount devices. As the use of surface mount devices increases, thermal mismatches and failures become likely at elevated temperatures. Thermal analyses often suggest new ways to make the design more functional and reliable.

Figure 7 shows cycles to failure for three multilayer boards (MLB) and one double-sided rigid (DSR) printed wiring board with surface mount devices. The test conditions were 60 cycles per day of temperatures between −20° and +130°C. When the thermal mismatch is large, the solder joints fail in 20 to 40 cycles (see boards J and K).When the thermal mismatch is small, the solder joints survive more than 350 cycles (see board H).[23]

Evaluate the use of composites

Designers must consider many issues in selecting materials such as plastics, wood, ceramics, or composites. Several reference books are useful.[24] [25] Composites are an example for which the technology is rapidly expanding and for which designers must consider trade-offs. A composite is a material that is composed of a high-strength material and a high-toughness material that can be formed, laminated, or molded to replace

---

[22]Smoluk, George M. "Thermal Analysis: A New Key to Productivity." *Modern Plastics,* February 1989, pp. 67–73.

[23]Sherry, W. M. and Hall, P. M. "Materials, Structures, and Mechanics of Solder Joints for Surface-Mount Microelectronics." *Proceedings of the Third International Conference on Interconnection Technology in Electronics.* Fellbach, West Germany, February 18–20, 1986, pp. 47–81.

[24]Young, James F. and Shane, Robert S., Eds. *Materials and Processes, Part A: Materials.* 3rd ed. New York: Marcel Dekker, 1985.

[25]Young, James F. and Shane, Robert S., Eds. *Materials and Processes, Part B: Processes.* 3rd ed. New York: Marcel Dekker, 1985.

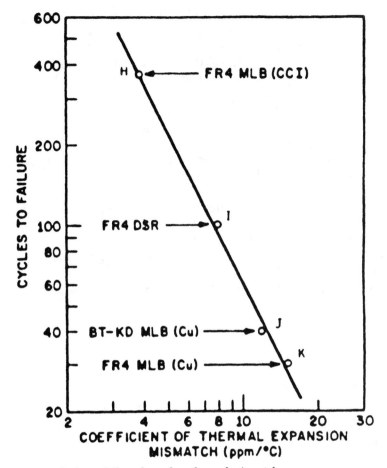

Figure 7 Cycles to failure depend on thermal mismatches.

a metal part. The advantages of composites are reduced weight, better corrosion resistance, better high-temperature resistance, and reduced life-cycle costs.[26]

Composites are used to provide:

- electromagnetic shielding
- electrical and thermal conductivity
- reflective surfaces

---

[26]Daane, John H., Horwath, John A., and Miller, Harold L. "New Materials" in *Manufacturing High Technology Handbook*. Eds. Donatas Tijunelis and Keith E. McKee. New York: Marcel Dekker, 1987, pp. 411–456.

Today, composites comprise about 14% of the structural weight of commercial and military aircraft.[27] For example, McDonnell Douglas used composites for the horizontal stabilizer on the F-14 aircraft and for the tail assembly or empennage on the F-15 aircraft. Bell Helicopter uses composites for its helicopter blades. The skin of Northrop's B-2 bomber is made entirely of composites to provide the stealth characteristics as well as rigidity and weight savings.

Composites may require new methods of processing and fabrication. Automated techniques may be needed to reduce the manufacturing costs. For example, a wing made of aluminum can be manufactured by one worker while a wing made of composites may require 20 workers. There are significant process problems in joining and cutting composites. Key questions are how to handle them, form them, and automate their manufacture.[28] Trade-off studies may show that advantages of composites including fuel economy, temperature resistance, and reduced life-cycle costs may outweigh the higher material and manufacturing costs.

When considering composites or any material, a designer should ask:[29]

- Will the product be safe?
- Can existing tooling be used?
- Are new processes required?
- Are new quality-assurance tools or techniques required?
- Were any weaknesses detected during design simulations?

Devise an overall strategy for selecting parts

Strategies for designers to use in selecting high-quality parts while reducing parts proliferation include:

- select parts that meet requirements for functionality, reliability, producibility, design strategy, and cost
- select preferred parts from approved databases

---

[27]"Materials: Backbone of Aerospace Designs." *Design News,* April 9, 1990, pp. 25-28.

[28]Owen, Jean V. "Assessing New Technologies." *Manufacturing Engineering,* June 1989, pp. 69–73.

[29]Katz, Harry S. and Brandmaier, Harold E. "Concise Fundamentals of Fiber-Reinforced Composites" in *Handbook of Reinforcements for Plastics.* Eds. John V. Milewski and Harry S. Katz. New York: Van Nostrand Reinhold, 1987, pp. 6–13.

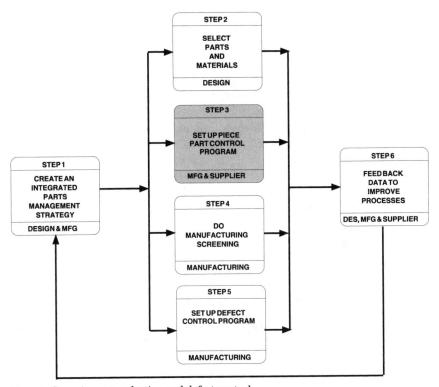

Figure 8 Steps in parts selection and defect control.

- follow procedures for obtaining approval if a needed part is not in the database
- ensure stresses are less than maximum rating
- choose the appropriate derating percentage (e.g., use derating curves)
- ensure derating is cost-effective
- do trade-off studies to decide which material to use (e.g., whether to use composites or metal parts)

Step 3: Set Up a Piece Part Control Program

Qualify and certify suppliers

Suppliers and customers must agree that the quality of the suppliers' parts matches what the customer expects. Customers expect a defect rate of 100 parts per million (ppm) or less for integrated circuits and a defect rate of 10 ppm or less for discrete components. One way to bring about customer confidence is with supplier-manufacturer alliances. These alliances are recent and still evolving. Increasingly, manufacturers want to work with suppliers who are willing to learn and change in order to control the quality and reliability of their piece parts.

The automobile industry, notably Ford Motor Co., has developed supplier-manufacturer alliances that produced cost-effective products of high quality. These programs resulted in an improved supplier base, long-term contracts for the suppliers, and improved operations for both the supplier and the customer. Implementing programs like these will lead to significant improvements in defense acquisition.

Supplier-manufacturer alliances based on cooperation and mutual benefit are becoming increasingly popular because they:

- allow quality standards to be applied consistently

- reduce the rescreening of microcircuits that meet requirements for high quality and reliability

- reduce the number of similar parts and thus costs of parts proliferation

- reduce costs by making just-in-time manufacturing possible (e.g., smaller inventories, lower carrying charges, less downtime)

With supplier-alliances, designers can select parts from proven suppliers chosen for their component quality and reliability, timeliness of delivery, and financial soundness.

Supplier alliances, reduced rescreening, and just-in-time manufacturing are important even with the priority ratings that earmark the first available product for the military contract. The alternative strategy of relying on rescreening adds the risk of delays and slipped schedules due to rejected lots.

Cost of ownership.   Many companies select suppliers on a cost-of-ownership basis (e.g., cost impact of rejects, late deliveries, receiving inspections) rather than just the cost of the part.[30] To do this, they consider:

- measurements, including assessment, monitoring, and reassessment

---

[30]Capitano, J. L. and Feinstein, J. H. "Environmental Stress Screening (ESS) Demonstrates Its Value in the Field." *Proceedings of the Annual Reliability and Maintainability Symposium,* 1986, pp. 31–35.

- attributes, including quality, reliability, manufacturing, delivery, service, and cost
- records, including defect rate, mean time between failures, and yield

Suppliers' process improvement. A key ingredient in fostering understanding between part suppliers and part users occurs when the users share their experience with the suppliers. Many assembly manufacturers also work with their suppliers to improve the suppliers' processes. The assembly manufacturers:

- provide requirements and specifications
- periodically review products and process controls
- help suppliers implement and use statistical process controls (SPC)

Achieving near-zero defects per million parts requires a stable process with all subprocesses in control. The goal is to discover the true sources of variation in the design or the manufacturing process.[31]

SPC must be applied effectively. Control charts should be used to monitor the ongoing process. Any deviations will then be quickly apparent. Deviations should trigger action to correct and eliminate the cause of the deviation. SPC is discussed in more detail in Step 5.

Select parts from qualified manufacturers list (QML). To reduce the cost of electronic components, semiconductor manufacturers and the government are participating in a joint effort to qualify the manufacturing process, under MIL-I-38535 for monolithic microcircuits and MIL-H-38534/MIL-STD-1772 for hybrids. The QML effort takes advantage of the near-zero defect levels many semiconductor suppliers achieve: incoming component defect levels below 100 parts per million (ppm). The QML program recognizes that incoming inspections are costly and not needed when suppliers implement statistical process control (SPC) programs to ensure quality is built in. The objective of QML is a 10-fold to a 100-fold decrease in the price of silicon microcircuits.[32]

With QML, the manufacturing processes are certified rather than individual parts as in the current Qualified Parts List (QPL) and MIL-M-38510, General Specification for Microcircuits. The traditional approach

---

[31]Maass, Richard. *World Class Quality: An Innovative Prescription for Survival.* Milwaukee, WI: ASQC Quality Press, 1988.

[32]Burgess, Lisa. "Thomas: Pushing the Pentagon Toward QML." *Military and Aerospace Electronics,* February 1990, pp. 39–40.

of certifying parts under the Joint Army Navy (JAN) programs and under the MIL-M-38510 program was costly, lengthy, and inefficient.[33] Parts became obsolete almost as soon as they were qualified. The lengthy audit process had to be repeated with each upgrade. QML will eliminate the need to requalify parts from a certified line.

The challenge for the parts industry is to make QML work. For QML to be successful:

- top management must support the process

- manufacturing must understand the entire manufacturing process

- customers must give feedback

The key element of QML is statistical process control. This in-process monitoring of the manufacturing processes is used to ensure device yield and reliability. Two innovative evaluation features are the standard evaluation circuit and the technology characterization vehicle. The standard evaluation circuit is used to demonstrate the reliability and quality that result from the processes. It is designed to stress the worst-case geometric and electrical design rules and to allow easy diagnosis of failures.

The technology characterization vehicle contains test structures that monitor intrinsic reliability failure mechanisms such as electromigration, time-dependent dielectric breakdown, and hot carrier aging.

The benefits of the QML program are:

- better control of the manufacturing process

- better use of facilities

- fewer government audits and lower qualification costs

- predictable part costs

- improved delivery schedules

- earlier use of advanced commercial integrated circuit technologies in military systems

AT&T's facility in Allentown, PA was the first manufacturing facility to have its processes certified by the government. With QML certification, they can produce their metal-oxide semiconductor integrated circuits for the military on the same line as those for commercial purposes.

---

[33]Gardner, Fred. "Hold Down Ballooning Costs and Boost Quality." *Electronics Purchasing,* June 1988, p. 57.

Commit to continual process improvement. It is wise to select at least two suppliers for each part who are willing to join in supplier-manufacturer alliances. These suppliers commit to work towards world-class quality of their parts and materials. This commitment means that designers, suppliers, and manufacturing engineers aim for continual process improvement. The designer selects approved and verified materials. The designer also specifies tolerances appropriate to the operating and environmental conditions. The tolerances should not be too tight (difficult to manufacture) nor too broad (cause instabilities and malfunctions). The manufacturing engineer spots abnormal distributions and then designs experiments to find the root causes and appropriate solutions.[34]

Even though the supplier's SPC ensures parts of high quality and reliability, the OEM should have processes in place that provide early warnings of a quality or reliability problem and provide immediate feedback to the supplier. For example, instead of inspecting for an acceptable quality level, the OEM can verify the supplier's test results using standard sampling techniques.[35]

Developing alternative sources may be expensive and difficult. It is difficult, for example, to divide small-quantity orders between two suppliers. Companies should do trade-off studies to balance the benefits with the risks of single sources.

Tailor contracts for reduced rescreening. Many prime contractors have tailored their government contracts to reduce or eliminate rescreening. Litton Guidance, for example, accepts Texas Instruments parts without rescreening them because of the history of near-zero defects. Texas Instruments certifies that its defect rate is less than 100 ppm and shares its SPC data with Litton.

Standardize military drawings and numbering system. To improve quality and reduce costs, the Defense Electronics Supply Center (DESC) in Dayton, Ohio set up a program in which suppliers or assembly manufacturers generate standard military drawings (SMD) for standard military parts. This program was set up to stop the expensive proliferation of source control drawings (SCD). In the standard military drawings program, DESC lists suppliers on the SMD whose parts meet

---

[34]Maass, Richard. *World Class Quality: An Innovative Prescription for Survival.* Milwaukee, WI: ASQC Quality Press, 1988, p. 24.

[35]Maass, Richard. *World Class Quality: An Innovative Prescription for Survival.* Milwaukee, WI: ASQC Quality Press, 1988, p. 4.

the specification limits. Assembly manufacturers can generate additional SMDs for parts that are not listed.

DESC has also set up a new numbering system for military semiconductor products to reduce paperwork and errors. Each part will now have one 15-character number. The only variable is a letter that tells users, suppliers, and manufacturers what standards the part is designed to meet—B for JAN Class B devices, S for JAN Class S devices, M for SMDs, and Q or V for devices from the Qualified Manufacturer List (QML).[36]

Manage sources

Companies must decide on the best way to eliminate marginal devices. The ultimate goal is to select parts only from qualified and certified suppliers. To learn more about the supplier's quality and processes, companies may use source inspection or receive inspection temporarily until the data show that the supplier's processes are stable and producing parts with near-zero defects.

In source inspection, companies send their own inspectors to the supplier's site. This practice often has advantages over receiving inspection. Advantages include saving time, equipment and labor, not having to return defective parts, and reducing the failures at higher assembly levels. Also, the inspectors learn about the supplier's practices on-site.

Source inspection may be ineffective if there is a strained relationship between the supplier and the assembly manufacturer or if the source inspector interrupts process flows. Source inspection may be preferable to receiving inspection, but it is less preferable to supplier-manufacturer cooperative alliances, which aim to build in quality through process controls. In these alliances, the manufacturer and supplier jointly share data and improve the processes.

If the supplier fully inspects and screens parts within a stable process-control environment, the assembly manufacturer can:

- eliminate duplicate test equipment
- reduce inventories due to the high percentage of usable parts
- use resources better
- pass along lower costs to the end user

Corporations such as Texas Instruments, General Dynamics, AT&T and many others have instituted supplier programs to reduce incoming in-

---

[36]Keller, John. "One Part, One Number: DESC Simplifies IC Buys." *Military and Aerospace Electronics,* May 1990, p. 1.

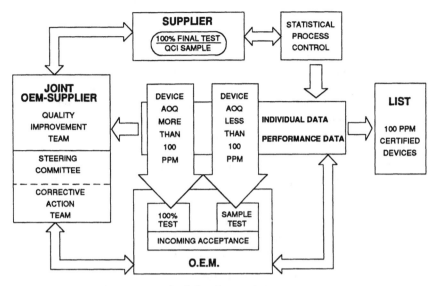

Figure 9 OEm-supplier strategy for defect-free parts.

spection costs. In today's environment of automation, high rework costs, and reduced inventories, companies try to avoid incoming inspections to monitor the supplier's quality. Today, OEMs and suppliers are working together to solve problems. These teams try to correct problems at the source rather than relying on incoming inspection.

Figure 9 shows how Texas Instruments and some of its suppliers are achieving defect-free parts without unnecessary rescreening. (Note that AOQ in the figure means average outgoing quality.)[37]

Use effective screening

Screening ideally occurs as part of the original manufacture of the part. In screening, stress is applied to devices to find marginal devices. Ideally, a failure mode analysis team uses the results of the screening to find root causes of the marginal devices and how to prevent them.

Screening helps the supplier, manufacturer, and customer. The benefits include:

- fewer failures at higher assembly levels and in the field

- less rework and repair on assemblies using screened parts

---

[37]Bindhammer, Carl and Krog, John. "An Electrical Test Correlation Experience." *Integrated Circuit Screening Report,* Institute of Environmental Science, November 1988, pp. 4-5.

- improved customer satisfaction
- reduced warranty or field failure costs
- feedback on the supplier's process-control system and more confidence in the supplier's components and material

With an effective screening program, the percentage of marginal devices should decrease to near zero. When this zero-defect level is reached and sufficient process control in manufacturing is achieved, companies may decide to reduce the screening or replace it with sampling.[38] This decision to reduce the screening depends on the technology, the cost of screening, how the system will be applied, and the contractual obligations.

Evaluate piece part quality and reliability

Ensuring piece part quality and reliability is vital. During the 1980s, there were divergent perceptions of the quality and reliability of military parts. The defect rate, which was actually quite high in the early 1980s, improved in the middle 1980s when many suppliers upgraded and standardized their screening and worked toward continuous improvement of their processes. By the mid 1980s, suppliers were reporting an average defect rate of about 1,000 parts per million (ppm). Assembly manufacturers and the military, however, perceived the average defect rate to be an order of magnitude higher, about 10,000 ppm.[39]

Even with the supplier improvements, the assembly manufacturers' and the military's perceptions of poor quality persisted. Strained relationships were common. Suppliers blamed assembly manufacturers for inaccurate testing, and assembly manufacturers blamed suppliers for poor quality. Many DoD contracts required 100% rescreening of incoming components in spite of the improvements in quality.[40]

Root causes for the defects.   There were many reasons for these different perceptions. The data did include parts with inadequate quality and limited performance and environmental capabilities that were improperly used in military systems. And, in many cases, the

---

[38]Klinger, David J., Nakada, Yoshinao, and Menendez, Maria, Eds. *AT&T Reliability Manual.* New York: Van Nostrand Reinhold, 1990, p. 52.

[39]Golshan, Shahin and Oxford, David B. "ESSEH Parts Committee Overview." *Integrated Circuit Screening Report,* Institute of Environmental Sciences, November 1988, p. 3-1.

[40]Oxford, David B. "Total Quality Management: Business Aspects and Implementation." *Integrated Circuit Screening Report,* Institute of Environmental Science, November 1988, p. 2-1.

assembly manufacturers and the suppliers used different equipment and screening methods. But perhaps the main cause of the different perceptions was that the average defect rate was misleading.

During 1986, for example, Texas Instruments' incoming-inspection data showed an average defect level of about 4,000 ppm. This average was computed by calculating the total confirmed rejected parts divided by the total parts screened and then normalizing to one million. This method of calculating the average, however, obscured key information. The 4,000 ppm average could result if each part type had a defect rate of about 4,000 ppm or if a few part types had a very high defect rate and most part types had a lower defect rate. The average would be misleading in that case.

Preliminary analyses at Texas Instruments showed that the average was indeed misleading: 31% of the part types contributed to the high defect rate, but 69% of the part types were defect-free when they left the supplier's site.

Figure 10 shows these data.[41]

Of the 31% part types that were responsible for the high defect rate, analog devices (e.g., linear interface, control, and converter circuits) showed the most correlation failures. That is, the assembly manufacturer and the supplier obtained different results. Correlation problems are likely with analog devices because it is harder for the supplier and the assembly manufacturer to obtain the same results screening analog

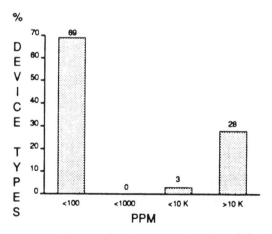

Figure 10 Percent of device types at various defect rates.

[41]Bindhammer, Carl and Krog, John. "An Electrical Test Correlation Experience." *Integrated Circuit Screening Report,* Institute of Environmental Science, November 1988, p. 4-1.

devices than digital devices, even with the same equipment and screening methods. The Texas Instrument analyses also showed that some parts were damaged during rescreening due to mishandling, insertion errors, and electrical overstress.[42]

In 1986, to investigate this correlation problem, the Institute of Environmental Sciences formed an ESSEH Semiconductor Parts subcommittee with representatives from suppliers, assembly manufacturers, test labs, and the military. This subcommittee operated within a larger committee called the Environmental Stress Screening of Electronic Hardware (ESSEH) committee.

The subcommittee set up a pilot program of three pairs of suppliers and original-equipment manufacturers (OEMs) to analyze the integrated-circuit screening data and resolve the divergent perceptions. In early 1986, one pair, Texas Instruments Defense Electronics Equipment Group (the OEM) and Texas Instruments Semiconductor Group (the supplier), formed joint OEM-supplier quality improvement teams to focus corrective action on the part types with high defect rates.

As a result, Texas Instruments increased the percentage of near-zero defect part types from 69% in 1986 to 95% by 1987.[43] This improvement, shown in Figure 11, resulted from failure analyses done by the OEM-supplier quality improvement teams. Their failure analyses helped prevent defects due to overstress and poor handling.

Conclusions. After studying the failure-analysis data from the three pairs in the pilot program, the ESSEH Semiconductor Parts subcommittee concluded:

- rescreening does not improve the quality or reliability of a device and in many cases may reduce the reliability of the device by inducing additional failures (e.g., from wrong insertions, electrical overstress, poor handling)

- rescreening should be avoided when the data supplied to the assembly manufacturer indicate a stable process that results in near-zero defects

- test procedures and manufacturing processes at the suppliers' facility should be periodically reviewed for effectiveness

---

[42]Golshan, Shahin and Oxford, David B. "ESSEH Parts Committee Overview." *Integrated Circuit Screening Report,* Institute of Environmental Sciences, November 1988, p. 3-1.

[43]Bindhammer, Carl and Krog, John. "An Electrical Test Correlation Experience." *Integrated Circuit Screening Report,* Institute of Environmental Science, November 1988, p. 4-4.

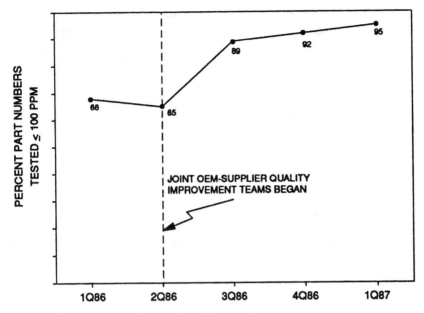

Figure 11   Improved percentages of defect-free part types.

- handling procedures should be reviewed for possible damage from electrostatic discharge and other problems
- customer-supplier teams are effective in solving problems

Screen at the piece part level

Screening at the piece part level is usually cost-effective for the supplier. Finding and removing defective parts early before they are shipped to the assembly manufacturer pays off. The supplier usually gets more business with a reputation for quality and reliability. The assembly manufacturer is spared the expense and delays of rescreening or failures at higher assembly levels.

Examples of screening at the piece part level include particle impact noise detection, high-temperature burn-in, highly accelerated stress testing, thermal shock, and hermeticity. Step 4 describes environmental stress screening that is applied to boards, subassemblies, and finished systems.

By definition, screening is done on 100% of the devices. Whether to use sampling or screening depends on the type of defects expected or the customer's requirements. For random defects (e.g., bonding defects), 100% screening is needed. For defects related to the particular manufacturing batch (e.g., defective metallization due to masking defects), sampling may be effective. Sampling tests the manufacturing process, 100% screening tests the components.[44]

Particle impact noise detection (PIND). For some device types (e.g., open-cavity devices, metal cans), a test of particulate contamination is particle impact noise detection (PIND), which can detect microscopic particles. To find loose particles in a sealed package using the PIND test, the packaged device is shaken with highly sensitive acoustic monitors attached to it.

Screening is more difficult for particles that are not actually loose but have the potential to come loose during mechanical or other operational stresses. These failures are often unexpected, unrepeatable, and hard to screen out. They usually come from defects and flaws that have been aggravated during manufacture.[45]

PIND has been controversial for the last ten years. Some government agencies, NASA for example, are strong advocates. Others question its usefulness, saying it is better to correct the process that is causing the loose particles rather than trying to screen for them. When improvements in the process result in near-zero defects, PIND may be no longer cost-effective.

High-temperature burn-in. Before suppliers ship the devices to assembly manufacturers, they expose devices to elevated temperatures while all electrical connections are exercised. The purpose of this burn-in is to find marginal devices that would otherwise fail in the assembled equipment. Suppliers analyze devices that fail burn-in and use the results to improve their design or manufacturing processes.

In the typical burn-in process, devices experience temperatures of 125°C for at least 48 hours.[46] High-temperature tests effectively screen microchemical flaws (e.g., contamination) in semiconductors and also weak bonds due to heel cracking. Some failure mechanisms (e.g., oxide defects) may not be sensitive to temperature but may be accelerated by overvoltage, overcurrent, or other stress. Tests that combine elevated

[44]Amerasekera, E. A. and Campbell, D. S. *Failure Mechanisms in Semiconductor Devices.* New York: John Wiley and Sons, 1987, p. 114.

[45]Amerasekera, E. A. and Campbell, D. S. *Failure Mechanisms in Semiconductor Devices.* New York: John Wiley and Sons, 1987, p. 112.

[46]Buck, Carl N. "Improving Reliability." *Quality,* February 1990, pp. 58–60.

temperature with mechanical stress are usually the most effective screens.[47] This combined thermal and mechanical screening is not yet frequently used.

Highly accelerated stress testing. Many semiconductor manufacturers are using highly accelerated stress testing (HAST) to stimulate corrosion failures in plastic-encapsulated devices. The devices are exposed to high temperature and humidity under pressure, which greatly accelerates wear and forces failures due to corrosion. This new accelerated approach takes less time than the traditional approach of exposing devices to 85°C and 85% relative humidity at ambient pressure for at least 1,000 hours. The traditional approach, called 85/85 testing, requires several weeks of testing. With HAST testing, however, the devices are exposed to 120°C and 85% relative humidity for only 100 hours.

Many studies have shown correlations between 85/85 testing at 1,000 hours and 120/85 testing (HAST) at 100 hours. Intel, for example, found that the failure mechanisms that HAST finds at the wafer level are similar to those found in actual packaged devices. Intel stressed the need to keep the test environment free from contamination to make sure the failure data reflect fabrication problems only. Intel also pointed out the need for careful handling during HAST and traditional testing to avoid contamination and corrosion from chlorine and salts.

Even though the HAST test equipment is more costly, the shorter testing times make it cost-effective. HAST has also pointed out ways to improve reliability and reduce defects from contamination, corrosion, and handling.[48]

Thermal shock. Thermal shock exposes a component or system to a rapid change in temperatures over a specified range at a determined rate to stimulate latent defects into failure. Thermal shock uses a liquid-to-liquid medium to provide the severe temperature shock environment when transferring between the temperature extremes. For example, thermal shock may subject devices to temperatures of −65°C and +125°C for about 10 seconds at each level. Thermal shock detects crystal defects and other packaging defects. Thermal shock is a cost-effective method of screening components and assembled circuit boards. It may be used at higher levels of assembly but the cost of the

---

[47]Amerasekera, E. A. and Campbell, D. S. *Failure Mechanisms in Semiconductor Devices.* New York: John Wiley and Sons, 1987, p. 104.

[48]Comerford, Richard. "Turning up the Heat on Stress Testing." *Electronics Test,* January 1990, pp. 20–23.

monitoring equipment may be high. It may be destructive and should be used cautiously.[49]

Mechanical shock. Shock tests determine the device's relative resistance to damage when dropped. The shock level is measured in multiples of gravity (G's).[50]

Temperature cycling. Temperature cycling screens for manufacturing defects (e.g., wire-bond defects, poor package seals, cracked dies). In a typical test, devices are put in hot/cold air-to-air cycling chambers for 10 cycles of $-65°C$ to $+150°C$ with 10-minute cycles.[51]

In temperature cycling, the temperature gradually increases or gradually decreases. In thermal shock tests, the temperature changes rapidly from one extreme to another.

Destructive physical analysis. Destructive physical analysis (DPA) may be used on prototypes, models, and samples to look for defects and identify corrective action. Because it is destructive, DPA is done on a small, representative sample. DPA might include x-ray tests, hermeticity residual gas analysis, particle impact noise detection, and internal disassembling to check for wire bonding, die-attach quality, and internal wire construction. On complex or submicron processes, a scanning electron microscope might be used.

Hermeticity. Hermeticity screens check for broken or cracked package seals. These hermetic seals control the gas leakage over the device's life cycle. Fine and gross leak tests may be used to stimulate these types of latent failures. Fine and gross leak tests are both necessary to detect different magnitudes of leak rates. Two fine leak tests are often used: helium leak testing and krypton gas testing. The most commonly used gross leak test is the fluorocarbon bubble test. For more details, see the Institute of Environmental Sciences' guidelines.[52]

---

[49]Amerasekera, E. A. and Campbell, D. S. *Failure Mechanisms in Semiconductor Devices.* New York: John Wiley and Sons, 1987, p. 105.

[50]Sanders, Robert T. and Green, Kent C. "Proper Packaging Enhances Productivity and Quality." *Material Handling,* August 1989, pp. 51–55.

[51]Amerasekera, E. A. and Campbell, D. S. *Failure Mechanisms in Semiconductor Devices.* New York: John Wiley and Sons, 1987, p. 106.

[52]Institute of Environmental Sciences. *Environmental Stress Screening Guidelines for Parts.* Mount Prospect, IL: Institute of Environmental Sciences, September 1985, p. 2-9.

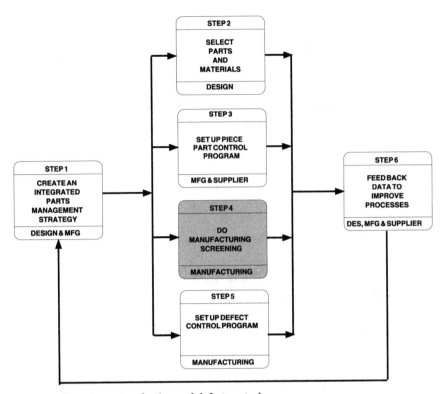

Figure 12 Steps in parts selection and defect control.

Step 4: Do Manufacturing Screening

Environmental stress screening (ESS) consists of exposing a circuit card, unit, or system to an environment that will induce latent defects to fail. These failures are *latent defects* because they are not detected through standard test methods, but they will emerge over time.

Screens to consider are thermal shock, temperature cycling, and random vibration. Screens may be used in combination to stimulate failures, e.g., temperature cycling and random vibration. The key factor with combination screens is whether the added stresses will expose significantly more failures and justify the cost.

Failures found by ESS fall into two categories:

- workmanship errors
- process defects

Define ESS objectives

The objective of environmental stress screening is to establish that the manufacturing process is in control and capable of manufacturing the product. Even a qualified process will produce a few products that are outside the acceptable limits, because of random variations within the process itself. It is better to find and eliminate these latent defects in the factory than in the field.

Develop plans for ESS

Because screening is expensive, planning is needed to obtain the greatest benefit at the lowest cost and at the least risk to good units. Planning helps:[53]

- select screens that stimulate the likely flaw without damaging good units (e.g., temperature cycling, random vibration, thermal shock, high-temperature burn-in)

- select the appropriate level of intensity, duration, temperature range or rate, vibration, repetitions, etc.

- use ESS at the lowest manufacturing level to find workmanship and assembly defects when rework is least expensive

- select the appropriate screen for technology being used (e.g., through-hole vs. surface-mount)

- demonstrate the screens' cost-effectiveness

- decrease ESS when the manufacturing processes are in control and the defect rate is too low to make ESS cost-effective

Select screen levels. Engineers must choose levels carefully to find as many flaws as possible without false positives (non-defects) or false negatives (missed defects) and with minimal damage to the product. Choosing correct levels is an iterative process. First, a low-level screen is run. If that does not stimulate latent defects, a higher level is run. If the higher level damages the unit, the engineers may decide to lower the level or modify the design.[54]

Five methods are in use, for example, for selecting the appropriate level for random vibration screening. Table 5 presents advantages and disadvantages of the five methods. For more details, see the Institute of Environmental Sciences guidelines.[55]

---

[53]Hobbs, G. K. "Development of Stress Screens." *Proceedings of the Annual Reliability and Maintainability Symposium,* 1987, pp. 115–119.

[54]Hobbs, G. K. "Development of Stress Screens." *Proceedings of the Annual Reliability and Maintainability Symposium,* 1987, pp. 115–119.

[55]Institute of Environmental Sciences. *Environmental Stress Screening Guidelines for Assemblies.* Mount Prospect, IL: Institute of Environmental Sciences, March 1990, pp. 37–41.

TABLE 5   Advantages and Disadvantages of Methods to Determine Screening
Level for Random Vibration

Methods	Advantages	Disadvantages
Tailored Spectral Response: Uses flaw pre-cipitation threshold to de-velop input spectrum and tailor screening level	■ Only method able to develop spectral char-acteristics ■ Least likely to damage good hardware ■ Shortest vibration ex-posure during develop-ment	■ Needs spectral analysis equipment and skilled operators ■ More expensive and time-consuming than tai-lored overall response method ■ May not be effective for new technology
Tailored Overall Response: Uses overall internal response levels to develop screening level	■ Similar to tailored spectral response method but less expen-sive and complex	■ Unable to adjust spectral characteristics
Step-Stress Tests: Sets the screening level be-tween the operating level and one-half the design or tolerance limits	■ Straightforward empir-ical method useful for existing and developing technology ■ Defines item design limits and makes equipment stronger	■ Some risk of overstress if design limits are un-known ■ Design may have to be changed to make it stronger
Fault-Replication Tests: Increases screening level until seeded (i.e., known) faults are replicated	■ Supplements step-stress tests	■ Hardware may not have replicable failure modes and it may be hard to seed hardware faults re-alistically
Heritage Screen: Derives screening level from past experience	■ Minimum development resources required and thus easier to obtain resources	■ Transparent dissimilari-ties may make screen in-adequate or damaging

ESS can be used successfully if the engineers carefully calculate stress levels to avoid degrading the equipment for normal use.[56] However, the long-term effects of ESS may limit a product's life. If this is likely, simu-lation may be used to find potential failures.

Choose screens.   An effective screen stimulates likely latent flaws without damaging good products. No one screen will find every type of flaw. Some flaws are stimulated by thermal cycling, some by vibration, and some by voltage cycling. Some flaws can be stimulated by several

[56]RADC. *Stress Screening of Electronic Hardware*. RADC Technical Report TR-82-27. Rome Air Development Center, 1982.

screens. Ignoring these flaw-stimulus relationships would lessen the effectiveness of environmental stress screening.

It is important to distinguish between latent or hidden defects and actual defects. Latent defects, which arise from irregularities in manufacturing processes or materials, become actual defects when exposed to environmental stimuli.[57] Effective screens find these latent defects without causing additional harm. Table 6 illustrates characteristics of effective and ineffective screens.[58]

Verify screens. One way to verify a screen is to seed the hardware, that is, to put known defects into the hardware. Effective screens will find most of these defects; ineffective screens will not.[59] It is not always possible to seed realistically.

Decide when to use ESS

ESS can be useful at the circuit-card, unit, or system level. In general, the cost per failure is lowest if ESS is used at the lowest possible level, but the detection efficiency is best at the highest levels because interface errors as well as other types of errors can be detected. Each project must decide which level is most cost-effective. Table 7 presents some advantages and disadvantages of ESS at the circuit-card, unit, and system levels.[60]

TABLE 6　Characteristics of Useful and Poor Screens

Useful Screens	Poor Screens
Precipitate flaws quickly	Fail to stimulate latent flaws
Stimulate adequate proportion of latent-defects	Induce additional defects
Apply accelerated stress but not overstress beyond the equipment's design limits	Consume too much of equipment's operating life

---

[57]Hobbs, G. K. "Development of Stress Screens." *Proceedings of the Annual Reliability and Maintainability Symposium,* 1987, pp. 115–119.

[58]Department of the Navy. *Navy Manufacturing Screening Program.* NAVMAT P-9492, May 1979.

[59]Hobbs, G. K. "Development of Stress Screens." *Proceedings of the Annual Reliability and Maintainability Symposium,* 1987, pp. 115–119.

[60]Fuqua, Norman B. "Environmental Stress Screening." Paper presented at the Joint Government-Industry Conference on Test and Reliability, AT&T Bell Laboratories, May 4, 1990.

TABLE 7   Advantages and Disadvantages of ESS at Different Levels

Level	Advantages	Disadvantages
Circuit-Card	Cost per failure is lowest Small size and mass permit batch screening and fast rates of temperature change	Detection efficiency is relatively low Interface errors are not detected
Unit	Can power and monitor performance during screen Detection efficiency is higher than at circuit-card level Interconnections (e.g., backplanes) are screened	Large mass precludes fast rates of change without costly equipment Cost per failure is higher than at circuit-card level Temperature range is lower than at circuit-card level
System	All potential sources of failures are screened Unit interconnections are screened Detection efficiency is highest	Screening at temperature extremes is difficult and costly Large mass precludes vibration screens without costly equipment

Implement environmental stress screening

Examples of environmental stress screening include temperature cycling and random vibration.

Temperature cycling. Temperature cycling is widely used to find defects in electronic equipment including circuit cards, assemblies, and systems. It is used during development to find and eliminate design problems and during production to find and eliminate defective units, processes, and workmanship. With temperature cycling, the temperature gradually increases and then gradually decreases for a set of predetermined times. The rate of change of temperatures depends on the specific heat of the unit or system being screened.

How many temperature cycles are needed depends on the complexity. Test and failure rate data show that six cycles are adequate for equipment with about 2,000 parts; 10 cycles are needed for equipment with about 4,000 or more parts. When unscreened parts are used, more than 10 cycles may be needed.

Temperature ranges of −65°F to +131°F are common. The maximum safe range of component temperature and the fastest rate of change of hardware temperatures will give the best screening. The optimal rate of change depends on the size and mass of the hardware.[61]

---

[61]Department of the Navy. *Navy Manufacturing Screening Program.* NAVMAT P-9492, May 1979, pp. 5–7.

Examples of defects screened stimulated by temperature cycling include poor solder joints, welds, and seals; shorted wire turns and cabling due to damage or improper assembly; fractures, cracks, and nicks in materials due to unsatisfactory processing.[62]

Temperature cycling is an accepted procedure for testing the reliability of surface-mount attachments under accelerated conditions. Fatigue and relaxation-type mechanisms commonly cause solder-joint failures in the field. Many fatigue failures are due to cyclic thermal variations that cause the component and the substrate to expand differently.

Random vibration.   Random vibration is appropriate when the design is sufficiently mature, i.e., no major unsolved design problems exist and no major design changes are expected. Random vibration, which excites many modes at once, has been found to be more effective than the single-frequency sine test. Using an apparatus with 100 simulated defects, Grumman compared random vibration with sine fixed-frequency and sine sweep tests at different levels for varying amounts of time. Random vibration was the most effective. It found faults the other tests missed. Eliminating these faults can prevent degradation in reliability.

Damaging vibration may occur during product use, handling, or transporting. To determine a product's susceptibility to vibration, a product is subjected to randomly changing frequencies (usually from 2 to 200 cycles per second). A critical component may resonate, fatigue, and then fail. Failure is likely if the vibrating range that damages the unit is the same as the vibrating range of the truck, train, or airplane that transports the unit. If the ranges are the same, cushioning or dampening is needed.[63] Examples of latent failures stimulated by random vibration include poor solder joints, poor connections, improperly seated connectors, and improperly mounted components.

Use ESS results for prevention

The effectiveness of a screen must be evaluated with factory and field failure rates and the screen parameters adjusted accordingly. ESS re-

---

[62]Department of the Navy. *Navy Manufacturing Screening Program.* NAVMAT P-9492, May 1979, pp. 5–7.

[63]Sanders, Robert T. and Green, Kent C. "Proper Packaging Enhances Productivity and Quality." *Material Handling,* August 1989, pp. 51–55.

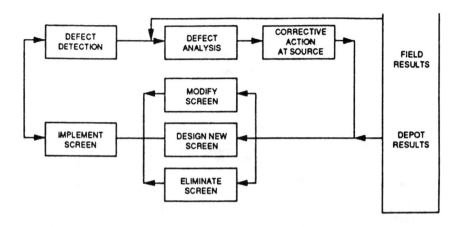

CLOSED-LOOP SYSTEM

Figure 13 ESS as a closed-loop system.

sults can help find root causes and corrective actions to prevent defects.[64] The purpose is to remove the sources of the defects. Many companies use feedback from ESS results to improve their design and manufacturing process. Fault rates are lower, as well as scrap and rework. Figure 13 illustrates the dynamic, closed-loop feature of an effective ESS program.[65]

With corrective action in place, the defects should fall to near zero. If the near-zero trend continues and the team is confident that the sources of defects have been removed, the team may decide to decrease or eliminate that screening and perhaps concentrate on other areas with greater sources of defects.

Step 5: Set Up a Defect Control Program

Control and prevent defects

Screening components at the suppliers' site will reduce component defects in assembled circuit boards. But defects originate during the as-

[64]Department of the Navy. *Best Practices: How to Avoid Surprises in the World's Most Complicated Technical Process*. NAVSO P-6071, March 1986, pp. 6-51 to 6-54.

[65]Institute of Environmental Sciences. *Environmental Stress Screening Guidelines for Assemblies*. Mount Prospect, IL: Institute of Environmental Sciences, March 1990, p. 6.

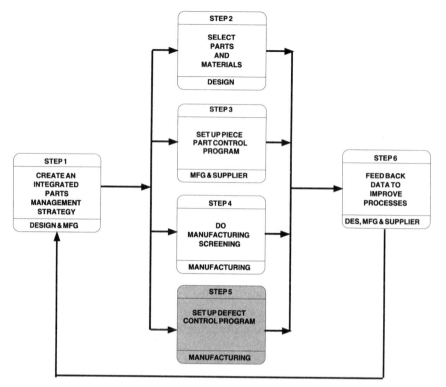

Figure 14   Steps in parts selection and defect control.

sembly process through bending and cutting of component leads, over-heating during soldering, and electrostatic discharge. Thus, the assembly process should be analyzed to find the root causes of defects.[66]
  Below are strategies for defect control:

- use robust design to optimize the design, manufacturing, and assembly processes

- use statistical process control to reduce variation in all processes and prevent defects

[66]Amerasekera, E. A. and Campbell, D. S. *Failure Mechanisms in Semiconductor Devices*. New York: John Wiley and Sons, 1987, p. 113.

- use a closed-loop defect control system
- use automated tracking and automated reporting for yield and defects
- use failure mode analysis (FMA) to find root causes and prevent recurring defects in components and in the manufacturing process

Use robust design

The goal of robust design is to make designs less susceptible to variations in materials, processes, or environmental conditions. An example of a robust-design technique is *design of experiments,* which engineers use to improve design and manufacturing processes.[67] For example, the techniques of Genichi Taguchi, an internationally known quality expert, combine statistical methods with engineering.

Advocates of robust design point out that the effort to reduce product failures in the field will also reduce the number of defective products in the factory. Systems that are designed to withstand wide variations in actual use can better withstand variations in factory processes and conditions. *Design to Reduce Technical Risk* has more details on the principles and application of robust design.

Use statistical process control

Statistical process control techniques are relatively simple tools for process monitoring, problem solving, and communication. When managers and employees use the statistical tools properly, they generate the continual improvements that prevent defects.

The logic of using statistical process control techniques is:

- Variability in processes causes many defects.
- Variability can be analyzed using statistical methods.

Teach basic SPC techniques.   Use of the statistical methods promotes better communication. Graphs, for example, clarify data and stimulate discussion on which processes are working well and which need to be improved. People can more readily see patterns and identify problems. A

---

[67]Phadke, Madhav S. *Quality Engineering Using Robust Design.* Englewood Cliffs, NJ: Prentice Hall, 1989.

small amount of data displayed graphically gives a lot of information about the process and the potential sources of defects. People can discuss them objectively and decide the best strategies for improvement. More information on SPC concepts and techniques is available in a number of reference books.[68] [69] [70] [71]

Effective companies teach basic SPC techniques to all employees, including machine operators and other shop-floor people. These techniques require familiarity with simple mathematical calculations, not statistical theory. The techniques are best taught with as many everyday terms as possible, with job aids for the step-by-step procedures, and with examples based on realistic situations.

Below are some useful SPC techniques that many companies teach to all employees:

- use check sheets to collect data

- use histograms to describe and analyze a process

- use Pareto analysis to separate the "significant few" from the "trivial many" and thus identify the key improvement opportunities

- do cause-and-effect analysis to find root causes of problems (e.g., people, methods, materials, equipment, and environment)

- construct control charts, including mean and range (X-bar and R) charts

- use control charts to identify which processes are acceptable and which processes are producing a high rate of defects (e.g., P charts, which show percentages of defective parts, and C charts, which show counts of the defects)

Figure 15 shows how SPC techniques can be used together to collect data, identify defect causes, and assess whether the improvements are stabilizing the process and preventing defects.[72]

---

[68]Aft, Lawrence S. *Quality Improvement Using Statistical Process Control.* New York: Harcourt Brace Jovanovich, 1988.

[69]Coppola, Anthony. *Basic Training in TQM Analysis Techniques.* Griffiss Air Force Base, NY: Rome Air Development Center, 1989.

[70]Juran, J. M. and Gryna, Frank, Eds. *Juran's Quality Control Handbook.* 4th ed. New York: McGraw-Hill, 1988.

[71]Kane, Victor E. *Defect Prevention: Use of Simple Statistical Tools.* New York: Marcel Dekker, 1989.

[72]Kane, Victor E. *Defect Prevention: Use of Simple Statistical Tools.* New York: Marcel Dekker, 1989, p. 354.

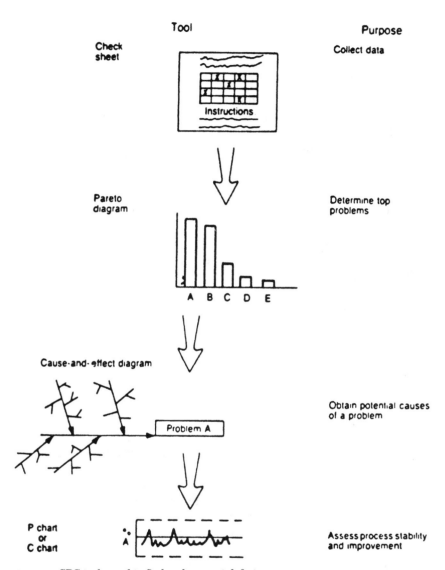

Figure 15 SPC tools used to find and prevent defects.

Teach advanced SPC techniques.   Advanced SPC techniques help engineers and other technical staff reduce variation, and thus defects, in products and processes. These techniques include use of multivariate charts to identify the major sources of variation, design of experiments to verify solutions, and capability charts to find opportunities for continual improvement. The capability charts include indices useful for process engineers, machine operators, and managers.

Use a closed-loop defect control program

In a closed-loop program, the objective is to prevent mistakes and defects with real-time process monitoring rather than detect and document defects afterwards. Information on defects and yields goes to the people whose activity affects performance. This information must be timely. Otherwise the system is a defect correction system rather than a defect prevention system. With timely information, corrective action can and should occur in real time. Each employee must be trained in SPC techniques to interpret the real-time results and immediately measure the effect of the corrective action on the process.

Honeywell's defect-reduction process. Honeywell, for example, is implementing a closed-loop system to eliminate the "hidden factory" of extensive rework in its torpedo operations. They establish teams, each with a production engineer, a quality engineer, and a factory representative. The teams meet weekly to discuss yield and defect data. Managers from the three areas also attend. The weekly meetings promote a regular and disciplined look at factory performance.

To prepare for the weekly formal review, the operators and engineers review the data daily. They use process control charts to measure process stability and to show immediately the effectiveness of corrective action.

In the weekly meetings, the teams select targets, set goals, interpret data, identify causes of significant defects, prioritize and implement solutions, and measure and communicate results. They use Pareto analysis to select targets with consistently high defects.

The teams establish a time-phased plan of continuous improvements. For example, Figure 16 shows improvements in defects per unit as a result of a series of corrective actions from 1980 to 1987 at Honeywell. (The "Apple Orchard" in the figure is an automated data collection system.)[73]

The closed-loop nature of the system helps improve the processes continually. With timely and accurate information, the teams focus on the areas with the lowest yields and the highest number of defects. As these areas improve, others receive attention.

Figure 17 compares actual defects per unit with the goals.[74]

---

[73]Raasch, Daniel. *The Defect Reduction Program.* Hopkins, MN: Honeywell, 1985.

[74]Raasch, Daniel. *The Defect Reduction Program.* Hopkins, MN: Honeywell, 1985.

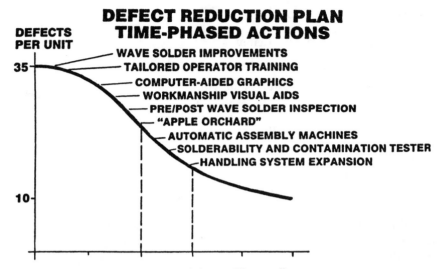

Figure 16   Corrective actions reduce defects at Honeywell.

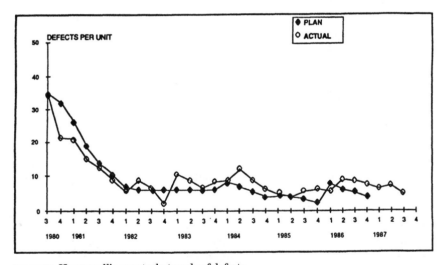

Figure 17   Honeywell's quarterly trends of defects.

Among the keys to success are:

- management actively participates
- shop-floor employees help recognize and solve problems as they occur
- a designated leader organizes, directs, and monitors
- data are collected accurately and timely
- process improvement occurs continually

Track yields and defect rates

Many companies have instituted automated systems to help track yield, which is the percentage of usable boards, units, or systems. The systems can also track removal rate by part, lot, supplier, and date. The yield and defect data are entered into databases that can provide daily, weekly, or monthly reports and trends for different products, manufacturing lines, and suppliers. With these data, engineers can work on low-yield, high-defect-rate areas.

The key feature of automated systems is automatic identification and tracking of components, units, and systems. Companies are using advanced technologies to enter data directly into computer systems, thus avoiding the time and errors of manual entry.[75] Table 8 describes some of these technologies.

For automated systems to facilitate data reduction, correlation, and trend analysis, it is important to make sure they have inherent controls and safeguards. Information such as data, part number, serial number, and test location is entered into the database system automatically. Often, however, trouble symptoms, troubleshooting results, and rework is entered manually. This manual entry can cause the database to be incomplete or inaccurate. Database entries must be reviewed and corrected often to ensure the failure data are accurate and complete.

Analysis of the failure data often suggests corrective actions to improve the yield and decrease the defects. These actions may include increasing the frequency of process checks, notifying suppliers of high defect rates, changing the test programs, and making reports available sooner. Table 9 gives examples of problems and solutions that may emerge after a failure analysis team investigates defect data.

Effective companies graph the yield and defect-rate data and post them prominently to make the data visible to each employee. Employees

---

[75]Beckert, Beverly, Knill, Bernie, and Rohan, Thomas. "Integrated Manufacturing: New Wizards of Management." *Automation,* April 1990, pp. 1–26.

TABLE 8 Technologies for Automatic Identification of Parts, Units, and Systems

Technology	Definition	Comments
Bar code scanning	Machine-readable code of bars and spaces	Fast, accurate, and inexpensive data entry
Radio frequency	Transponder located on the object being tracked operates on a unique frequency	Requires antenna to pick up and deliver signal to host and reader to interpret the distinctive data stream; electrical noise may cause problems
Optical character recognition	Stylized numbers and letters recognizable by scanner that transmits data to the computer	Characters can be read by people and machines
Voice data entry	Operator speaks into a microphone using a preprogrammed vocabulary	Frees operator's hands and avoids typing errors
Machine vision	Optical scanning of identification labels, objects, or documents	Imaging process is more complex and more expensive than bar codes and optical character recognition

TABLE 9 Examples of Problems and Solutions Found with Defect Analysis

Problem	Solutions
High failure rate for a particular capacitor	Examine the design and the component, tell supplier about component failures
False indications from test sets	Clean test probes, clean connectors, check calibration, check power supply
Test points hard to probe	Make test points accessible, use low solids flux to prevent flux build-up on component side, clean test points, increase pressure
Bent pins	Investigate handling and assembly processes

at all levels learn how to interpret graphs and control charts in classes on statistical process control techniques. Continual improvements in processes and products come about when employees use data to classify the type and cause of defects. Failure mode analysis is used to find the root causes of problems and likely solutions.

Use failure mode analysis

Failure data often indicate which components failed on which boards, the location of the removed components, and the quantity removed. These data on removal rates and defect rates, however, may not be meaningful without failure mode analysis (FMA). FMA can determine the root causes of the high removal rate. Is the removed component actually defective? Is the diagnostic method effective and efficient? Is the test procedure properly applied? Is the test set miscalibrated? Or is the component being damaged by poor handling?

When designers and engineers have access to FMA data, they can monitor trends and investigate problems early to prevent scrap, rework, repairs, and line shutdowns.

FMA helps answer these questions:

- How can a product or process fail? Potential failure modes in a manufacturing process include machines out of alignment, inconsistent temperatures, etc.

- What will happen if a product or process does fail? Effects range from minor delays to major failures to catastrophic safety issues.

- How can design and manufacturing prevent these failures? Teams set priorities based on how severe the effect is, how likely it is, and whether they can find a way to prevent it.

Use effective FMA techniques.   Typically, FMA begins at the early design stages and continues throughout the production process. Effective companies form FMA teams of designers, manufacturing engineers, quality engineers, reliability engineers, and others who can evaluate data on the design and the processes and combine their different skills and experience to analyze and solve problems quickly. FMA teams identify types of failures and find ways to prevent them, using the following techniques:

Reproduce the failure.   Reproducing the failure is essential in order to vary the stimulus conditions and obtain additional data on the sources of the problem.

Determine the root cause.   The root cause is the original event that triggered the problem. Extensive testing is often needed, especially if the failure occurs intermittently at a low rate. To find root causes, teams ask "why" repeatedly until they arrive at the underlying cause of the failure. Teams begin by looking for common problems. In manufacturing, for example, common problems include bending and cutting of component leads, overheating during soldering, electrostatic discharge, etc.

Identify the failure mechanism. The team must determine the entire failure mechanism to find cost-effective solutions and ways to prevent future problems. It is important to find the entire failure mechanism because there may be several failure mechanisms activated at different stress levels.

Verify the failure mode analysis. After the team identifies the failure mechanisms, the team chooses the most cost-effective and feasible solution. The next step is to test the solution under stress, including the worst-case configurations. The team verifies the root cause with tests that show the failure occurs when the failure mechanism is activated and does not occur when the failure mechanism is not activated. When the solution is verified, the corrective action can be implemented.

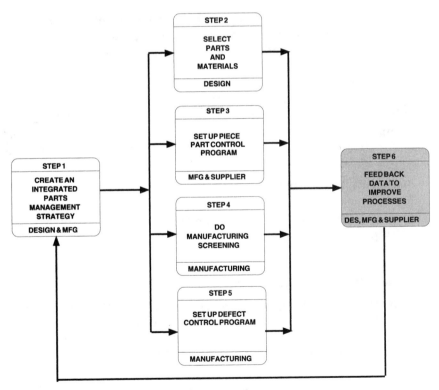

Figure 18   Steps in parts selection and defect control.

Step 6: Feed Back Data to Improve Processes

Obtain useful feedback information

Use of feedback leads to improved processes and systems. With feedback, designers, suppliers, and manufacturers can see what is going well and what needs to be improved. Effective feedback gives information to those who can correct or change the product or process. Ineffective feedback results when information goes only to people who have little power to correct and prevent the problem.

Feedback is most helpful when it improves processes and prevents problems in the future. It is less helpful when it is just used in short-term fire-fighting. Prevention is the key to successful feedback.

Feedback should be presented in a usable form. Key information should not be buried in lengthy, hard-to-read reports that require readers to extract and summarize the useful information in separate reports. In general, it is more efficient for the report preparers to abstract and summarize the metrics and information for the users of the data.

Feedback should be timely, rather than arriving months after the action that caused the problem. Whether the appropriate interval is daily, weekly, or monthly depends on the volume, the number of defects, and the delivery schedules.

To set up useful feedback systems, teams must:

- identify feedback flows (e.g., who, what, when, and how much)

- set up an integrated system to collect, analyze, and report the feedback (e.g., extract and summarize data such as which components are the source of several trouble reports)

- institutionalize the use of feedback to fix defects quickly and prevent future problems (e.g., use feedback to update the design rules of the corporate computer-aided design system or alter the manufacturing process)

Consider information flows

Information on design requirements, customer specifications, and special needs should be given to designers, manufacturers, and suppliers.

Feedback information should flow back and forth among suppliers, designers, and assembly manufacturers. Examples of useful information include:

- defect rates and yield categorized by product, component, and supplier
- results of FMA analyses

- periodic reports on supplier performance (e.g., quality, reliability, delivery, service, and cost)
- changes in manufacturing processes or assembly equipment

Designers need appropriate feedback from manufacturing. If feedback is available in a database or in a few timely reports, it can be used to improve designs and the design process.

With feedback from the assembly manufacturer, the supplier can do root cause analysis to see what is causing the defects. For example, some defects may originate at the assembly location due to mishandling, electrical overstress, or reverse insertion during retesting. Sharing the results of the analysis may lead to improved processes at the supplier and the assembly location.

Consider many sources of feedback information

Factories have data from test results, failure mode analyses, and all stages of manufacturing. For example, circuit boards that fail in-circuit or functional tests are analyzed to find the cause. Trends of the defect rates for the past month, quarter, and year can provide useful feedback to suppliers and can be used to upgrade the component and supplier databases.

Useful feedback also comes from field operations. Below are sources of field feedback and possible results of analyzing that feedback. (For more information, see AT&T's *Performance Limit Reference Guide* which includes the Field Feedback template.)

- Warranty and Repair Feedback—What are the root causes and corrective actions for damaged and non-operative systems? Which units are often returned for repair?

- Installation and Maintenance Feedback—What parts are easily broken or damaged during shipping and installation? Are the cables and connectors the right length and configuration? Which units have more maintenance than expected?

- Customer Trouble Reports—What are the most frequent sources of problems? What percentage of the problems originate in design, manufacturing, or supply?

- Customer Surveys—Which customer needs are being met? Which needs are not being met?

- DoD Field Failure Return Program—What are reliability trends for parts in use?

Government information data exchange program (GIDEP). The Government Information Data Exchange Program (GIDEP) is a cooperative activity between government and industry to make maximum use of existing knowledge. GIDEP coordinates and distributes information that participants send to them. Reports are available on mechanical, electrical, electronic, hydraulic, and pneumatic devices. Information on failure experiences is continually collected on parts, materials, safety and health hazards, and test instrumentation. Information is indexed by subject, supplier, and part number. For critical items, the GIDEP Operations Centers issues an alert or a safe-alert to inform all participants immediately.

The four GIDEP information exchanges include:

■ failure experience data interchange, which has information on failure analysis, failure experiences, problems, and diminishing manufacturing sources

■ engineering data interchange, which distributes reports on the testing and qualification of parts, materials, and systems

■ reliability-maintainability data interchange, which has information on operational field performance, accelerated life testing, and reliability and maintainability tests on systems and equipment

■ metrology data interchange, which has calibration procedures, maintenance manuals, and measurement techniques

An organization can use the information to improve the reliability, maintainability, safety or cost of systems being designed, being manufactured, or in field use. Participants are alerted about potential failures and hazards so they can avoid costly errors, prevent malfunctions, and save resources.

For example, the June 1986 GIDEP Alert traced catastrophic electrical failures in circuit card assemblies to cracks in the glass frit used to seal microcircuit packages. Contamination entered through the cracks and caused corrosion. The cracks were caused by thermal shock as the microcircuits were dipped in hot solder and then into a cleaning solvent without cooling. For more details, contact the GIDEP Operations Center, Corona, CA.

Use feedback data to improve processes

Successful companies use feedback data to improve processes continually. They make sure the feedback information clarifies problems, rather than obscures them. For example, averages may be misleading and obscure the root causes of problems. For example, an average defect rate of

2,000 ppm could result even if 80% of the part families were defect-free and 20% of the part families had defect rates much higher than 2,000 ppm. It is important to have additional data to identify the problem and find root causes and solutions.

Here are some likely questions that feedback data will help answer:

- How can the designers' component selection process be improved?

- How can the supplier management process be improved?

- How can the suppliers' piece part control process be improved?

- How can the manufacturers' screening and defect control programs be improved?

The answers will help improve processes and update information in databases accessible to each participant in part selection and defect control.

Chapter

3

Application

In this chapter are examples of the use of the principles and processes discussed in the Procedures chapter. The examples include:

- McDonnell Douglas's integrated database
- Improving reliability in the F/A-18 Hornet
- AT&T's supplier-management program
- AT&T-supplier alliance on a widely used component
- Magnavox's supplier program
- Stress screening at Hughes Aircraft
- Monsanto's defect control program
- Motorola's six-sigma program

McDonnell Douglas's Integrated Database

To record, structure, and communicate information and to evaluate the downstream impact of design decisions, a diverse life-cycle team at McDonnell Douglas Helicopter uses an integrated database called the Integrated Design Environment—Aircraft (IDEA).[76] This common database collects and translates information from all the computer-aided tools the team uses. Figure 19 shows the interrelationships among engineering, producibility, supportability, cost, schedule, and accounting.

---

[76]Meyer, Stephen A. "Integrated Design Environment-Aircraft (IDEA): An Approach to Concurrent Engineering." Paper presented at the American Helicopter Society 46th Annual Forum and Technology Display, Washington, DC, May 21-23, 1990.

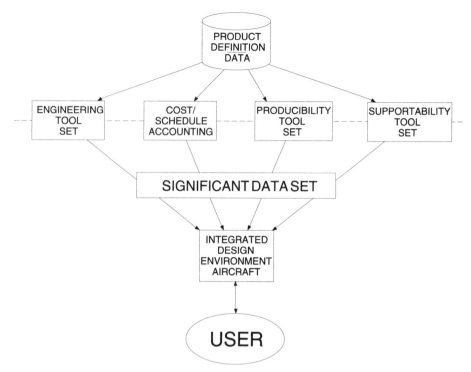

Figure 19   Schematic view of the IDEA integrated database.

Database characteristics

McDonnell Douglas Helicopter developed IDEA using an object-based, commercial database called the Integrated Development Environment. IDEA is object-based: its data structures are objects that make up complex networks. A landing gear picture, for example, can be used to get its definition and data.

IDEA is hierarchical—lower-level drawings relate to assemblies, assemblies relate to installations, installations relate to subsystems. An IDEA data object has many attributes and network associations. Figure 20 shows the data clusters for the LH aircraft system are subsystems or assemblies of parts that can be broken down into smaller parts and attributes.

Managing the data

Product definition data include system and subsystem specifications, drawings, process planning, technical documentation, supplier data, make-buy decisions, computer-aided design two-dimensional draw-

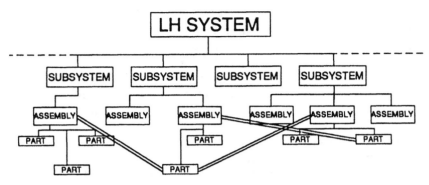

Figure 20 Hierarchical levels of data for the LH aircraft.

ings, weight and cost estimates, reliability and maintainability estimates, and lessons learned. Organizational tools manage design information and enhance the transition from development to production.
These tools associate product information and stocklists with the drawings to facilitate communication. The database has information on design criteria, manufacturing processes, and support methods.

Using the data

Different work groups can use the data for their individual needs. To
ensure the integrity of the data, everyone has access to the data in a
view-only mode. The ability to edit, move, or delete data from the
database is restricted by work group. All users can build flexible views
of the data for their personal use without compromising the data. IDEA
ensures that changes in one subsystem, for example, are communicated to all other affected groups.

This integrated database helps facilitate teamwork among people from
design, component engineering, reliability, maintainability, human factors engineering, and producibility. It allows product changes to be communicated rapidly to other members of the development team. It provides
efficient, paperless communication of design decisions.

Improving Reliability in the F/A-18 Hornet

Improving reliability with part selection, thermal analysis, and derating

The Navy fighter attack plane F/A-18 Hornet built in the late 1970s by
McDonnell Douglas was designed to meet performance and reliability
requirements three times more stringent than the fleet capability at

that time. Thus, the flight control electronics built by General Electric also had to meet stringent requirements. To do this, GE emphasized "reliability by design," with particular emphasis on parts selection and control, thermal analysis, and derating criteria.[77]

To improve reliability at the piece part level, they reviewed the circuits to minimize parts and ensure part quality. High-failure rate or high-usage parts were given 100% screening over the full temperature range ($-55°$C to $125°$C). To ensure the highest quality parts for microcircuits and discrete semiconductors, GE selected approved suppliers with histories of quality and timeliness.

GE used thermal analysis to define optimum piece part locations on the circuit boards and optimum circuit board locations within the unit for effective cooling. Heat sink rails on the circuit card were held in pressure contact with the unit by means of cam-action extractors. This design gave a positive path to remove heat from the circuit boards to the unit's cooling system.

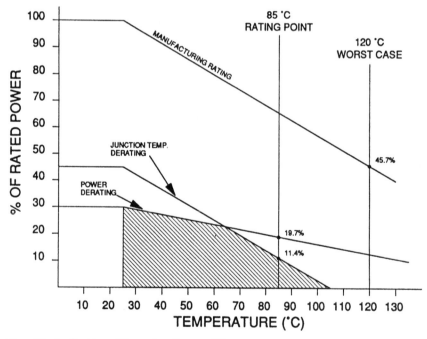

Figure 21 Application of three derating conditions.

[77]McGrath, J. D. and Freedman, R. J. "The New Look in Reliability—It Works." *Proceedings of the Annual Reliability and Maintainability Symposium*, 1981, pp. 304-309.

Figure 21 shows how derating characteristics were imposed on the devices. The figure uses a 2N6193 transistor operating curve. Each device had to meet three stringent derating criteria:

- operate at a maximum case temperature of 85°C
- maintain derated maximum junction temperatures on semiconductors
- remain within the manufacturer's rating at 120°C case temperature

The shaded area is the derated operating temperature range. The maximum percentage of rated power for this device is the intersection of the 85°C case temperature and the maximum junction temperature derating. To meet the three conditions, this transistor should not be operated at more than 11.4% of its rated power.

As the example shows, the operating power limit was often well below the manufacturer's rating. Exceptions were made for about 2% of the parts due to board area limitations and to avoid adding a significant number of integrated circuits. But overall, the flight control electronics met the stringent performance and reliability requirements.

AT&T's Supplier-Management Program

To improve quality and reduce costs, AT&T created a four-phase supplier-management program with selected suppliers. The four phases are: identify suppliers, evaluate their capabilities, select the suppliers who commit to near-zero ppm defect rate, manage the process to achieve continual improvements.

Phase 1: Identify needs

In this first phase, the goal is to identify needs of the customer as well as design and manufacturing. These needs are then translated into product requirements.

Phase 2: Evaluate suppliers

The quality, reliability, and manufacturing systems of potential suppliers are audited.

Phase 3: Select suppliers

Suppliers are selected after the evaluation data are reviewed by representatives from design, purchasing, quality, reliability, and manufacturing.

Phase 4: Manage and improve suppliers' process

AT&T collects quality data from its suppliers and analyzes and maintains the data in a central on-line database. Designers, customers, and suppliers have access to its quality data (defect rate measured in parts per million) and reliability data (measured in failures in time).

When a supplier has a record of near-zero defects for some time, AT&T no longer inspects that supplier's incoming parts. Instead, AT&T monitors the supplier's quality control. The monitoring becomes less and less intrusive as the quality becomes proven and more stable. The progression typically follows this sequence:

- 100% reinspection
- lot-by-lot acceptance sampling
- periodic review of products and product controls
- audit of process controls
- quality leadership programs and awards

The ultimate level is set using a total cost perspective that includes the cost of mission failure, repair, replacement, and lost opportunities. An undersea cable that costs $1,000,000 to $10,000,000 to repair will undergo more inspection and stress testing than a cable used on land.

The essence of the AT&T's supplier-management program is assess, monitor, and reassess quality, reliability, manufacturing processes, service, delivery, and price. Table 10 illustrates AT&T's supplier-management form. In the cells of the table would be the dates and results of the system audits during the three phases.

The initial assessment provides the baseline from which subsequent progress is measured. Joint supplier-manufacturer quality improvement plans result from monitoring. Reassessment and periodic updates offer assurance of continued improvement. Monitoring with

TABLE 10   AT&T's Supplier-Management Form

Element	Assess	Monitor	Reassess
Quality			
Reliability			
Manufacturing Processes			
Service			
Delivery			
Price			

statistical process control is encouraged. Data are mutually shared. Joint failure mode analysis teams are formed as needed to identify problems and find corrective action.

System audits occur during assessment of quality, reliability, manufacturing, service, delivery, and price. The quality system audits, for example, examine the control of measurement and test equipment, quality in specification and design, production control, and corrective action procedures. The reliability system audits examine the collection, computation, and use of reliability data, the failure mode analysis procedures, and the past reliability history. The manufacturing system audits examine the facilities, process controls, and problem resolution procedures.

During the monitoring phase, AT&T uses data from the system audits on the six elements (quality, reliability, manufacturing, service, delivery, and price) to classify the suppliers as Preferred, Reliability-Monitored, Acceptable, Restricted, or Unclassified. The goal is two preferred suppliers for each product family (e.g., capacitors, resistors, potentiometers, inductors). Preferred suppliers have a proven history of excellent quality and reliability, superior performance in service, plus competitive pricing.

Measure quantifiable results

With the database, engineers can continuously monitor results to make sure the program is meeting its goals. The program is designed to improve quality, provide conforming material, reduce variability, and reduce the number of suppliers. These goals are being met. Quality engineers can spend more time training and consulting, rather than inspecting.

AT&T-Supplier Alliance on a Widely Used Component

AT&T set up a supplier alliance on a widely used ceramic capacitor that reduced parts proliferation, inventory, and a 12-week ordering interval to one week. After a year, AT&T had zero defects in 24 million pieces. An integrated team with representatives from quality management and engineering, design, manufacturing, and purchasing achieved these results.

Work with the supplier

The team focused on a bypass capacitor that was used at least once in every circuit board. Previously, 30 different parts from five suppliers were used to accomplish the same function, with the same form and fit, making it difficult to manage quality and stock levels. The team used feedback from designers and manufacturing people to select the best two of the 30 parts: one through-hole device and one surface-mount device. The team then worked with the supplier to enhance and optimize the parts.

To reduce the 16-week lead time between the order and the delivery of the parts, AT&T set up an alliance with the supplier. AT&T shared yield data and forecasts, with actual equipment orders translated to part levels. The supplier agreed to ship full reels only with 5,000 parts per reel in increments of five reels per box to prevent errors and bad counts. AT&T adjusted its production to use box units of 25,000 parts. The sharing of data and supplier agreements allowed just-in-time manufacturing and eliminated counting errors. The supplier receives the forecasts each Monday and ships on Tuesday. The parts arrive at the storeroom on Thursday when the new production week begins.

Measure quantifiable results

To measure the effectiveness of the program, AT&T measured the number of parts with quality defects and the number of parts not on hand when needed. In the first year of the program, there were no missed deliveries and no quality defects in 24 million parts. (The process has been proven to be sensitive to seeded defects.) AT&T is now planning to apply these techniques to other mature parts and mature suppliers who can commit to high levels of quality, reliability, manufacturing, service, and delivery at competitive prices.

Magnavox's Supplier Program

Magnavox Electronic Systems worked with suppliers to cut rework costs by 80% and reduce lamination, interconnect fractures, and copper fractures on multilayer boards.[78]

To correct recurring problems in laminate separation and plated copper fractures, Magnavox began a program to learn more about their suppliers' capabilities. They first asked suppliers to complete a questionnaire about their measurement units, quality standards, and procedures. They reviewed the answers to find the most knowledgeable and capable suppliers whom they then visited to inspect their facilities.

---

[78]Gardner, Fred. "Magnavox's Message: Don't Settle for PC Board Garbage." *Electronics Purchasing,* March 1989, pp. 84-87.

Visit supplier facilities

On their plant visits to the multilayer-board producers, Magnavox evaluated the suppliers' capabilities. One of the items Magnavox looked for was whether the supplier had automatic optical inspection equipment. Optical inspection reduces the amount of scrap by helping pinpoint flawed laminates before the board is sandwiched together. Magnavox also looked for a functioning statistical process control (SPC) program, particularly in the areas of wet chemistry and plating. Daily charts of temperature, acidity, and impurity rates are keys to monitoring the process. Any discrepancies from specifications trigger corrective action. For example, too much tin may cause solderability problems. Another item was whether the printed circuit board templates are stored under correct temperature and humidity control. Most important, however, was whether quality is built in rather than bad material screened out.

Communicate expectations

Magnavox then gave a seminar for designated suppliers and Magnavox field inspectors. The goal of the seminar was to clarify specifications, describe expectations for supplier process audits, and establish clear accept-and-reject criteria for the circuit boards.

For example, a supplier must meet the following standards to qualify as a designated supplier with an effective SPC program:

- All process variables must be controlled unless specifically excluded. An example is charting the specific temperature range for an operation.

- Each variable must be charted for at least 25 measurements before evaluating its importance.

- After 25 measurements, a variable may be adjusted or dropped if SPC shows the variable does not affect the process.

- Each facility must have an SPC policy available for review at all times. The charted variables must be prominently displayed in the workplace.

- The SPC charts must demonstrate yields, problems, and impact. The data must be able to predict results and show whether the process is in control.

Measure quantifiable results

Magnavox's efforts paid off. They learned how to manage their suppliers. Dollars spent on scrap and rework of finished boards were reduced by $400,000 between 1987 and 1988.

Stress Screening at Hughes Aircraft

This case study illustrates stress screening of the AIM-54C Phoenix missile built by Hughes Aircraft.[79]

Analysis, strategy, and implementation

The Phoenix missiles use complex state-of-the-art hardware with demanding performance requirements. The missiles track six different targets as far away as 125 miles and varying from sea level to high altitudes. Two of its electronic units have 1,300 and 1,500 parts, respectively. The missiles must have a mean time between failure of 500 hours of flight conditions, even after being stored for extended periods of time. To achieve this, Hughes used a stress screening program to prevent operational failures and find defects at the lowest level of assembly.

The components undergo screening at their manufacturers' site. Circuit cards undergo functional and temperature tests. Units, which are collections of cards, undergo random vibration tests. Missile sections, which are collections of units, undergo temperature, voltage, shock, and random vibration testing. The Phoenix missile, which has four sections, undergoes two operational tests. Table 11 shows the stress screen tests from the lowest level (semiconductor) to the highest level (missile).

Measure quantifiable results

Evaluation of 3,000 tests showed that increasing the number of temperature cycles from 6 cycles to 20 cycles at the circuit-card or chassis level increased the percentage of defects uncovered from 30% to 60% of the possible temperature-related defects. Figure 22 shows that superimposing the temperature and voltage extreme tests with 20 cycles increases the percentage to 100%. Thus, the screen uncovers all of the temperature-related defects, which are about 90% of the total defects.

Hughes also analyzed the data to see if the percentage of defects the screens actually found varied by assembly technology (e.g., wave-soldering, hand wiring, point-to-point wiring). The screens were most effective with the point-to-point wiring technology and least effective

---

[79]Wong, C. L. and Zimmerman, R. L. "Stress Screening Can Benefit a Pipeline Requirement." *Proceedings of the Annual Reliability and Maintainability Symposium,* 1987, pp. 125-129.

TABLE 11   Stress Screens and Tests for the Phoenix Missile

Level	Stress Screens and Tests
Semiconductor	High and low temperature Particle impact noise detection
Hybrid	Bond pull Particle impact noise detection Functional tests
Chassis (Circuit Card)	Temperature cycling Functional tests
Unit	Temperature cycling Random vibration Functional tests before and after vibration tests
Guidance Section	Harmonization Temperature and voltage Shock Random vibration Functional tests before and after vibration tests
Control Section	Shock Random vibration Functional tests before and after vibration tests
Missile	System functional tests

with the hand-wiring technology. Most of the defects were workmanship defects in manufacturing rather than component or engineering defects.

Hughes then used random vibration to trigger nontemperature-related defects. They decided to double the unit vibration test time and start section-level testing in the factory because they were finding many vibration-triggered defects at the missile level. They also used two cycles of temperature and voltage extreme tests at the section level. Functional tests are done before and after the stress screening tests.

Report the benefits

These tests plus the two system tests of the missile were expected to reduce storage and flight problems to one third of the previous level. For example, the percentage of missiles out for repair should decrease from 22% to 7%. The number of missiles required to be in the inventory dropped from about 1,176 to 938 with substantial cost savings.

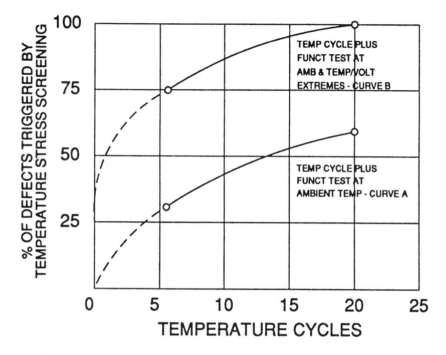

NOTES:

(1) CYCLES FROM -50 °C to +85 °C, 15 °C/MIN

(2) SAMPLE SIZE: OVER 600 CHASSIS

(3) TEMP/VOLT EXTREME: +71 °C, -55 °C

Figure 22   Adding temperature cycles and extreme tests.

Monsanto's Defect Control Program

Monsanto's defect control program boosted yields 35%, cut rejects 75%, and cut inventory 25%.[80] In the mid 1980s Monsanto's customers complained about poor service and inconsistent quality in the wafers Monsanto produced. In the past five years, however, Monsanto has established a model defect control program.

---

[80]Elliot, Marc. "Monsanto's Quality Turnaround." *Electronics Purchasing,* March 1989, pp. 95-98.

Reduce defects

To reduce the defects, they formed quality improvement teams with members from different disciplines. Some of the improvements reflect the input from several groups working together. For example, because the wafer-slicing department and the test and measurement department were far apart, it took several hours to find out if the slicing was accurate. The team proposed reconfiguring the equipment and layout. The new arrangement cut the cycle time from 50 hours to 6 hours.

They also improved the defect rate by closely monitoring their process. The statistical process control charts allow them to stay more closely within the optimum ranges. For example, one type of wafer used to require resistivity between 5 and 10 ohms per square even though the optimum resistivity is 7.5 ohms per square. They are narrowing the range closer and closer to the optimum.

Measure quantifiable results

Monsanto's defect rate went from 20,000 ppm to 1,500 ppm and is still improving. All 3,000 of its employees have been trained in statistical process improvement techniques and how to manage quality improvement as a daily part of their job.

Motorola's Six-Sigma Program

By 1992, Motorola wants to achieve a rigorous goal of "six-sigma quality" which is a defect rate of 3.4 parts per million units or 99.9997% accuracy.[81] Overall, defects have been cut from nearly 3,000 ppm in 1983 to less than 200 in 1989. Motorola was one of three companies to receive the first Malcolm Baldrige Award in 1988 on the strength of these improvements.

The term "six sigma" comes from the statistical distribution of variability around an average, as Figure 23 illustrates. In a normal distribution, 68% of the values fall within plus or minus one sigma or one standard deviation away from the mean, 99.7% fall within plus or minus three sigma or three standard deviations away from the mean. Plus or minus three sigma is 2,700 defects per million. 99.9997% of the values fall within plus or minus six sigma. To achieve six sigma or 3.4 defects per million, engineers must create more-robust designs (i.e.,

---

[81]Therrien, Lois. "The Rival Japan Respects." *Business Week*, November 13, 1989, pp. 108-118.

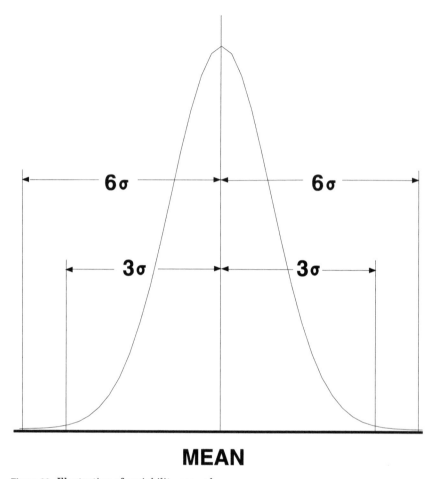

MEAN

Figure 23 Illustration of variability around an average.

those that tolerate more than the normal variation) or use tight control over the process to minimize variations from the optimal value.

Example

For example, a circuit board with a 10,000-ohm resistor can achieve a 3.4-ppm defect rate with robust design or with close control over the variation. A robust design would allow the board to operate within specifications even if the resistor varied about 5,000 ohms above or below the 10,000 average value. If such a robust design is not feasible, the variation of the resistors themselves must be kept within a narrow range.

Measure quantifiable results

To reach the six-sigma goal, Motorola needs a much higher level of quality in its suppliers and from their own workers. They are helping suppliers implement statistical process control and cross-disciplinary teams. Design, manufacturing, and marketing participate early in new-product development to ensure quality, producibility, and shorter cycle times.

The six-sigma goal of 3.4 ppm has already been reached for simple products like calculators and is the goal for all products for 1992. Measurements are used to check progress toward these tough goals. Each employee receives 40 hours of training on the principles of statistical process control linked directly to applications on the job. Cross-disciplinary teams often go through training together.[82]

---

[82]Haavind, Robert. "Motorola's Unique Problem: What to Do for an Encore." *Electronic Business*, October 16, 1989, pp. 60-66.

Chapter

4

Summary

This part has explained the best practices associated with informed decisions on parts selection and defect control.

The best practice that is key to minimizing the technical risks is setting up an integrated parts control program. This program should be far-reaching. Designers, manufacturing engineers, quality and reliability engineers, suppliers, purchasing representatives, and subcontractors should work together to meet customer needs. These groups are able to work more smoothly if they have common access to up-to-date information in integrated databases that share data, provide feedback, and foster continual improvements.

Best Practices for Each Template

The following are key best practices for the templates included in this part. When implemented, they will make an integrated parts selection and defect control program possible.

Parts and material selection template

A best practice in parts and material selection is a preferred-parts list to avoid a proliferation of nonstandard parts that vary in performance and reliability. In many companies, component managers and component-control committees help designers choose preferred parts suitable for the particular application.

Derating is useful to ensure the parts can withstand the stresses during use as well as during storage or transportation. Effective designers also use the results of stress analyses and thermal analyses in choosing parts and in locating them optimally in the circuits.

Piece part control template

A best practice in piece part control is a formal parts control program in which requirements flow down to subcontractors and suppliers. Another best practice is effective screening. Screening is most cost-effective when it is done at the supplier's site within a statistical process control program.

Successful companies look for root causes and corrective actions to prevent defects in the future. They actively manage their sources of supply by moving from receiving inspection, to source inspection, to supplier-manufacturer cooperative alliances that aim to build in quality through process controls.

In these alliances, the manufacturer and supplier jointly share data and work to improve the processes. When the data show the supplier's process is stable and producing near-zero defects, rescreening may not add value and may actually cause damage due to electrostatic discharge or electrical overstress.

Manufacturing screening template

A best practice in manufacturing screening is to understand environmental stress screening. Engineers need this understanding to choose screens that are strong enough to stimulate latent flaws in circuit cards, units, and systems and yet not too strong to cause unnecessary damage.

Another best practice is to tailor the screens to the particular process and technology. Temperature cycling and random vibration are useful in stimulating hidden assembly and workmanship defects.

Engineers also need to consider the appropriate assembly level. Screening done at lower assembly levels may not find latent defects, especially in interfaces and interactions among circuit cards and units. But screening done at higher assembly levels may fail to prevent expensive rework. Trade-off studies are needed to find the most appropriate assembly levels. Root cause analysis and corrective action should be incorporated into the screening program to prevent future defects.

Defect control template

Best practices in defect control focus on prevention, rather than on firefighting. Robust design techniques are used to optimize the design, manufacturing, and assembly processes. Statistical process control techniques are used to reduce variation in all processes and prevent defects. Automated systems are used to track and report yield and defects, thus avoiding the time and errors of manual entry. Failure mode analysis is used to find root causes and prevent recurring defects in components and in the manufacturing process.

Summary of the Procedures

Table 12 is a summary of the key practices in parts selection and defect control.

TABLE 12   Summary of the Procedures

Step 1: Create an Integrated Parts Management Strategy	
Procedures	Supporting Activities
Create integrated databases of preferred parts and forecasts, yields, and defects	■ Give database access to designers, manufacturers, subcontractors, project management, purchasing representatives, and supplier partners ■ Update information continuously to keep it current ■ Use data only to achieve mutual goals
Use series of linked databases	■ Include in each database information specific to it as well as index parameters to link databases to each other (e.g., specification number, supplier by location, part description or family)

Step 2: Select Parts and Materials	
Procedures	Supporting Activities
Establish preferred-parts lists	■ Include parts that meet criteria for performance, reliability, timely delivery, reasonable cost
Submit stocklists for approval	■ Make sure parts are approved ■ Audit parts for reliability, availability, manufacturability, and cost ■ Help designer select parts from equivalent components that meet precise specifications
Consider quality and reliability	■ Distinguish between quality (number of good parts that arrive at the next user) and reliability (ability to function for an expected time at an expected level) ■ Use reliability data in selecting components, for improving reliability, and for planning stress screening ■ Use MIL-HDBK-217 and other sources for component failure rates
Consider producibility during the design phase	■ Work with manufacturing engineers in selecting parts ■ Minimize number of parts ■ Select parts that are functional, reliable, available, and meet factory needs

TABLE 12   Summary of the Procedures (*Continued*)

Step 2: Select Parts and Materials	
Procedures	Supporting Activities
Consider derating	■ Include safety margins throughout the design ■ Use derating curves to show points of inflection where small increases in stress produce large increases in failure rates
Do stress analysis and thermal analysis	■ Use stress analysis and thermal analysis to measure stresses and temperatures and compare them to derating criteria
Evaluate use of composites	■ Consider advantages and disadvantages of composites in trade-off studies
Devise overall strategy for selecting parts	■ Select parts that meet requirements for functionality, reliability, producibility, design strategy, and cost ■ Use preferred parts that meet derating criteria
Step 3: Set Up a Piece Part Control Program	
Procedures	Supporting Activities
Qualify and certify suppliers	■ Ensure near-zero defect rates while reducing screening and rescreening that damages parts ■ Choose suppliers based on their record of quality, reliability, timeliness, and financial soundness ■ Work with suppliers to improve and stabilize their processes with statistical process control techniques ■ Select parts from suppliers on qualified manufacturers list (QML) ■ Use suppliers who commit to continual process improvement ■ Tailor contracts for reduced retesting ■ Standardize military drawings and numbering system
Manage sources	■ Decide on best approach to eliminate marginal devices (e.g., qualify and certify suppliers, source inspection, receiving inspection)
Use effective screening	■ Apply stress screening to devices to remove marginal devices ■ Find root causes of marginal devices ■ Reduce screening as marginal devices decrease to near zero

TABLE 12   Summary of the Procedures (*Continued*)

Step 3: Set Up a Piece Part Control Program	
Procedures	Supporting Activities
Evaluate piece part quality and reliability	■ Analyze defect rates to see which parts are defective ■ Find root causes for defects (e.g., mishandling, insertion errors, overstress, poor process controls) ■ Focus on corrective action for part types with high defect rates
Screen at piece part level	■ Consider advantages of particle impact noise detection to find loose particles vs. corrective action to eliminate sources of loose particles ■ Consider high-temperature burn-in to find contamination flaws ■ Consider highly accelerated stress testing to stimulate corrosion failures ■ Consider thermal shock to detect crystal defects and other packaging defects ■ Consider mechanical shock to determine device's resistance to damage when dropped ■ Consider temperature cycling to find manufacturing defects ■ Consider destructive physical analysis on prototypes, models, and samples to find defects and identify corrective ation ■ Consider hermeticity screens to find broken or cracked package seals

Step 4: Do Manufacturing Stress Screening	
Procedures	Supporting Activities
Define environmental stress screening (ESS) objectives	■ Improve reliability of circuit cards, units, and systems by stimulating factory failures rather than field failures
Develop plans for ESS	■ Select screens that stimulate flaws without damaging good units ■ Select appropriate level of intensity, duration, temperature range, vibration, repetitions, etc. ■ Choose effective screens ■ Demonstrate screens' effectiveness
Decide when to use ESS	■ Consider advantages and disadvantages of ESS at circuit-card, unit, and system level ■ Use ESS at lowest effective manufacturing level

TABLE 12   Summary of the Procedures (*Continued*)

Step 5: Set Up a Defect Control Program	
Procedures	Supporting Activities
Implement ESS	■ Consider temperature cycling to find defective units, poor solder joints, shorted wires due to improper assembly, and cracks due to unsatisfactory processing ■ Consider random vibration to find poor solder joints, poor connections, improperly seated and mounted components
Use ESS results for prevention	■ Use ESS results to find root causes and corrective actions to prevent defects ■ Decrease ESS as defects fall to near zero
Control and prevent defects	■ Analyze assembly process to find root cause of defects and strategies to prevent them
Use robust design techniques	■ Use robust design techniques to make design less susceptible to variations in materials, processes, or environmental conditions
Use statistical process control (SPC) techniques	■ Use SPC for process monitoring, problem solving, and communication ■ Teach basic SPC techniques to all: check sheets, histograms, Pareto analysis, cause-and-effect analysis, and control charts ■ Teach advanced SPC techniques to engineers and operators: multivariate charts, design of experiments, and capability charts
Use closed-loop defect control program	■ Prevent mistakes and defects with real-time process monitoring ■ Give information to people whose activity affects performance
Track yields and defect rates	■ Use bar code systems and computer databases to track yield and defects automatically ■ Analyze data to find frequent defects ■ Use results to find root causes and corrective actions for low-yield and high-defect rates
Use failure mode analysis (FMA)	■ Reproduce failure if possible ■ Determine root causes ■ Identify failure mechanisms ■ Verify failure mode analysis

TABLE 12 Summary of the Procedures (*Continued*)

Step 6: Feed Back Data to Improve Processes	
Procedures	Supporting Activities
Obtain useful feedback information	■ Feed back information to designers, manufacturers, and suppliers on part, circuit, and equipment failure rates
Consider information flows	■ Arrange for information to flow back and forth among designers, manufacturers, and suppliers ■ Feed back information to designers on circuits with too-tight tolerances ■ Feed back information to manufacturers on design requirements, customer specifications, changes in suppliers' processes, suppliers' failure rates ■ Feed back information to suppliers on results of FMA analyses and report card on supplier performance (quality, reliability, manufacturing, delivery, service, and cost)
Consider many sources of feedback information	■ Use factory FMA data and field data from installation, maintenance, warranties, trouble reports, and customer surveys ■ Use information from the Government Information Exchange Program (GIDEP)
Use feedback data to improve processes	■ Use data to improve component selection, supplier processes, stress screening, and defect control program

References

Aft, Lawrence S. *Quality Improvement Using Statistical Process Control.* New York: Harcourt Brace Jovanovich, 1988. Describes concepts and procedures for statistical process control, including Pareto analysis, cause-and-effect analysis, control charts, and capability charts. Includes case studies and examples.

Amerasekera, E. A. and Campbell, D. S. *Failure Mechanisms in Semiconductor Devices.* New York: John Wiley and Sons, 1987. Describes screening techniques for semiconductor defects including particle impact noise detection (PIND) for particulate contamination, high-temperature burn-in to eliminate device failures due to manufacturing defects, thermal shock to detect crystal defects and other packaging defects, temperature cycling to find structural defects (e.g., wire-bond defects, poor package seals, cracked dies). Discusses analysis of the entire manufacturing process to identify the root causes of defects. Discusses when to use sampling or 100% screening.

Anderson, R. T. *Reliability Design Handbook.* Chicago: IIT Research Institute, 1976. Provides design information, data, and guidelines for engineers to use to ensure a reliable end product. Describes methods for part control, derating, environmental resistance, redundancy, and design evaluation.

AT&T Performance Limit Testing Reference Guide. Describes government and industry best practices for design limit, life, and field feedback.

Ball, Graham. "How Well Do Materials Meet Designers' Real Needs?" *Materials: Proceedings of Materials Selection and Design.* London, England, July 1985, p. 208. Discusses the selection of materials on the basis of function, reliability, appearance, methods of manufacture, and cost.

Bannan, M. W. and Banghart, J. M. "Computer-Aided Stress Analysis of Digital Circuits." *Proceedings of the Annual Reliability and Maintainability Symposium,* 1985, pp. 217-223. Discusses stress analysis and derating to prevent stress-induced failures. Stress analysis ensures that the derating criteria are being met and verifies that the equipment can perform under worst-case conditions. Discusses manual and computer-aided methods of stress analysis.

Beckert, Beverly, Knill, Bernie, and Rohan, Thomas. "Integrated Manufacturing: New Wizards of Management." *Automation,* April 1990, pp. 1-26. Describes technologies to identify parts, units, and systems automatically and to track and analyze yields and defects.

Bindhammer, Carl and Krog, John. "An Electrical Test Correlation Experience." *Integrated Circuit Screening Report,* Institute of Environmental Science, November 1988, pp. 4-1 to 4-14. Discusses the reasons for the different perceptions of the quality and reliability of integrated circuits in the 1980s. Gives data that define the problem and the root causes. Describes improvements that resulted from joint supplier-manufacturer corrective action teams.

Bronikowski, Raymond J. *Managing the Engineering Design Function.* New York: Van Nostrand Reinhold, 1986. Discusses the design function and the design process, including parts and material selection.

Buck, Carl N. "Improving Reliability." *Quality,* February 1990, pp. 58-60. Describes the theory and practice of burn-in to screen for hidden failures in piece parts.

Burgess, Lisa. "Thomas: Pushing the Pentagon Toward QML." *Military and Aerospace Electronics.* February 1990, pp. 39-40. Describes the qualified manufacturers list (QML) which certifies manufacturing processes rather than individual parts.

Capitano, J. L. and Feinstein, J. H. "Environmental Stress Screening (ESS) Demonstrates Its Value in the Field." *Proceedings of the Annual Reliability and Maintainability Symposium,* 1986, pp. 31-35. States major cause of failures is due to defective component parts, not quality or design errors. Components are often defective because component manufacturers focus on decreasing unit cost rather than removing latent defects. Also, environmental test programs for components are often less stringent than the actual field environments.

Caruso, Henry. "Environmental Stress Screening: An Integration of Disciplines." *Proceedings of the Annual Reliability and Maintainability Symposium,* 1989, pp. 479-486. Discusses the distinct but complementary technical disciplines and backgrounds that are essential ingredients for developing and implementing an environmental stress screening program. Provides guidelines for developing consistent and effective environmental stress screening programs for electronic assemblies and systems.

Clech, J-P. and Augis, J. A. "Engineering Analysis of Thermal Cycling Accelerated Tests for Surface Mount Attachment Reliability Evaluation." Paper presented at the Seventh Annual International Electronics Packaging Society Conference, Boston, MA, November 8-11, 1987. Describes thermal cycling, which is an accepted procedure for testing the reliability of surface-mount attachments under accelerated life conditions. Discusses fatigue and relaxation-type mechanisms commonly responsible for solder-joint failures.

Comerford, Richard. "Turning up the Heat on Stress Testing." *Electronics Test.* January 1990, pp. 20-23. Describes a procedure called highly accelerated stress testing (HAST) to stimulate corrosion failures that takes less time than the traditional approach of exposing devices to 85°C and 85% relative humidity at ambient pressure for at least 1,000 hours. Describes the new approach in which devices are exposed to 120°C and 85% relative humidity for only 100 hours. States that the HAST failure-rate data are comparable to data found after 1,000 hours of traditional testing.

Coppola, Anthony. *Basic Training in TQM Analysis Techniques.* Griffiss Air Force Base, NY: Rome Air Development Center, 1989. Describes techniques for statistical process control which are useful in total quality management (TQM).

Daane, John H., Horwath, John A., and Miller, Harold L. "New Materials" in *Manufacturing High Technology Handbook.* Eds. Donatas Tijunelis and Keith E. McKee. New York: Marcel Dekker, 1987, pp. 411-456. Discusses physical characteristics, advantages and disadvantages, and applications for metal and fiber composites.

Department of Defense. *Total Quality Management Guide. Volume 1: A Guide to Implementation.* DoD 5000.51-G, January 1990. Provides a basic understanding of total quality management (TQM). Describes principles, tools, and techniques for continuous process improvement.

Department of Defense. *Transition from Development to Production.* DoD 4245.7-M, September 1985. Describes techniques for avoiding technical risks in 48 key areas or templates in funding, design, test, production, facilities, logistics, management, and transition plan.

Department of the Navy. *Best Practices: How to Avoid Surprises in the World's Most Complicated Technical Process.* NAVSO P-6071. March 1986. Discusses how to avoid traps and risks by implementing best practices for 48 areas or templates including parts and materials selection, piece part control, defect control, and manufacturing screening.

Department of the Navy. *Navy Manufacturing Screening Program.* NAVMAT P-9492, May 1979. Discusses characteristics of effective environmental screening programs including selecting the number of temperature cycles, the temperature range and rate of change, and the random vibration parameters.

Design to Reduce Technical Risk. Describes industry and government best practices for computer-aided design and computer-aided manufacturing; design policy, design process, and design analysis; and mission profile, trade studies, and design requirements.

Elliot, Marc. "Monsanto's Quality Turnaround." *Electronics Purchasing*. March 1989, pp. 95-98. Describes Monsanto's defect control program that increased yields, cut rejects, and cut inventory.

Finney, John W. "Pentagon, in Effort to Replace Army's Tanks, Finds Industry is Unwilling or Unable to Expand Production." *New York Times*, September 30, 1974, p. 12. Describes the pitfalls of relying on a single supplier who could not provide castings for tanks to replace those given away much faster than predicted.

Fuqua, Norman B. "Environmental Stress Screening." Paper presented at the Joint Government-Industry Conference on Test and Reliability, AT&T Bell Laboratories, May 4, 1990. Describes the advantages and disadvantages of environmental stress screening at different assembly levels.

Gardner, Fred. "Don't Settle for PC Board Garbage." *Electronics Purchasing*. March 1989, pp. 84-87. Describes how Magnavox worked with suppliers to cut rework costs by 80% and reduce lamination and interconnect separation and copper fractures.

Gardner, Fred. "Hold Down Ballooning Costs and Boost Quality." *Electronics Purchasing*, June 1988, p. 57. Compares the advantages of the qualified manufacturers list (QML), which certifies manufacturing process with the traditional parts certification method, which was costly, lengthy, and inefficient.

Garfield, Jerry and Bazovsky, Igor. "Economical Fault Isolation Analysis." *Proceedings of the Annual Reliability and Maintainability Symposium*, 1985, pp. 480-484. Discusses failure mode analysis at a radio-control development unit at Gould NAVCOM Systems Division.

Golshan, Shahin and Oxford, David B. "ESSEH Parts Committee Overview." *Integrated Circuit Screening Report*, Institute of Environmental Science, November 1988, pp. 3-1 to 3-6. Discusses the events that led to the formation of an ESSEH parts committee and joint supplier-manufacturer teams to isolate and solve quality problems in integrated circuits.

Guitard, Roger. "Reliability Data: A Practical View." *Microelectronics Reliability*, vol 29(3), 1989, pp. 405-413. Describes effective reliability data systems that include the failure mode, the failure cause, the corrective action, and the effectiveness of the corrective action. Describes how designers can then use the data to predict a device's reliability when it operates under similar conditions.

Haavind, Robert. "Motorola's Unique Problem: What to Do for an Encore." *Electronic Business*, October 16, 1989, pp. 60-66. Discusses Motorola's efforts at continued quality improvement after having won the prestigious Malcolm Baldrige award.

Heindenreich, Paul. "Supplier SPC Training: A Model Case." *Quality Progress*, July 1989, pp. 41-43. Describes how a supplier used Motorola's training materials to implement statistical process control (SPC) techniques at every level of the company.

Hnatek, Eugene R. *Integrated Circuit Quality and Reliability*. New York: Marcel Dekker, 1987. Discusses the quality and reliability considerations for integrated-circuit design. Includes material on the design and fabrication process, current technologies, sources of manufacturing error, and causes of failure and their remedies.

Hobbs, G. K. "Development of Stress Screens." *Proceedings of the Annual Reliability and Maintainability Symposium*, 1987, pp. 115-119. Discusses key aspects of environmental stress screening, including eliminating future defects by corrective action rather than defect detection and repair, selecting a screen that can stimulate likely flaws, selecting what level of stimulus to use with regard to intensity, duration, temperature range, vibration spectrum, and number of repetitions. Discusses how to prove a screen is effective and nondamaging to good products. Gives examples of defects screened out by temperature cycling.

Hyland, Charles and Shea, Joseph. "Environmental Stress Screening at Raytheon." *International Test and Evaluation Association Journal*, vol 9(3), 1988, pp. 32-37. Describes screening programs for two Raytheon programs including the screening environments and the results from factory testing and actual use.

"Incentive Payments to Concerns May End Tank Production Snag." *New York Times*, December 7, 1974, p. 58. Describes how the Pentagon found a second supplier and gave incentive payments for the two suppliers to increase their production of the castings needed for the Army tank.

Institute of Environmental Sciences. *Environmental Stress Screening Guidelines for Assemblies.* Mount Prospect, IL: Institute of Environmental Sciences, March, 1990. Describes concepts and procedures for random vibration and temperature cycling stress screening.

Institute of Environmental Sciences. *Environmental Stress Screening Guidelines for Parts.* Mount Prospect, IL: Institute of Environmental Sciences, September 1985. Describes concepts and procedures for stress screening for parts.

Juran, J. M. and Gryna, Frank, Eds. *Juran's Quality Control Handbook.* 4th ed. New York: McGraw-Hill, 1988. Describes the theory and procedures for basic statistical methods, for statistical process control, and for the design and analysis of experiments.

Kane, Victor E. *Defect Prevention: Use of Simple Statistical Tools.* New York: Marcel Dekker, 1989. Describes the concepts and tools needed to establish a defect prevention system for any work activity. Discusses statistical process control concepts and simple statistical tools useful in solving a variety of manufacturing or administrative problems.

Katz, Harry S. and Brandmaier, Harold E. "Concise Fundamentals of Fiber-Reinforced Composites" in *Handbook of Reinforcements for Plastics.* Eds. John V. Milewski and Harry S. Katz. New York: Van Nostrand Reinhold, 1987, pp. 6-13. Discusses the elements, properties, design, and testing of fiber-reinforced composites.

Keller, John. "One Part, One Number: DESC Simplifies IC Buys." *Military and Aerospace Electronics,* May 1990, p. 1. Describes the Defense Electronics Supply Center's new numbering system for semiconductors to reduce paperwork and errors.

Kidwell, George. "Aircraft Design—Performance Optimization." in *Engineering Design: Better Results Through Operations Research Methods.* Ed. Reuven Levary. New York: North-Holland, 1988, pp. 276-293. Discusses procedures to optimize aircraft design, including wind design, parts and material selection, and engine constraints.

Klinger, David J., Navada, Yoshinao, and Menendez, Maria, Eds. *AT&T Reliability Manual.* New York: Van Nostrand Reinhold, 1990. Discusses reliability concepts, device and system reliability, device hazard rates, and techniques to monitor reliability.

Maass, Richard. *World Class Quality: An Innovative Prescription for Survival.* Milwaukee, WI: ASQC Quality Press, 1988. Discusses the importance and benefits of improving component quality and reliability. Discusses supplier alliances and when incoming inspection is needed. Discusses techniques for achieving near-zero defects per million parts from a stable process with narrow tolerances with all subprocesses in control.

Malcolm, John G. "R&M 2000 Action Plan for Tactical Missiles." *Proceedings of the Annual Reliability and Maintainability Symposium,* 1988, pp. 86-92. Describes the results of an extended storage study on tactical missiles and the implications for stress screening.

Martini-Vvedensky, J. E. "Computer-aided Selection of Materials." *Materials: Proceedings of Materials Selection and Design,* London, July 1985, pp. 269-273. Discusses sources of information including structured handbooks that help the designer choose parts and materials. Also lists computerized data bases for plastics, metal alloys, ceramics, and other nonmetallics. These data bases include information from suppliers and the technical literature.

"Materials: Backbone of Aerospace Designs." *Design News,* April 9, 1990, pp. 25-28. Discusses benefits and applications of advanced composites which are now 14% of the structural weight of commercial and military aircraft.

McGrath, J. D. and Freedman, R. J. "The New Look in Reliability—It Works." *Proceedings of the Annual Reliability and Maintainability Symposium,* 1981, pp. 304-309. Discusses how General Electric improved the reliability of the F/A-18 Hornet's flight-control electronics. GE emphasized "reliability by design," with particular emphasis on parts selection and control, thermal analysis, and derating criteria. GE selected approved vendors with histories of quality and timeliness. Each part had to meet stringent derating criteria. GE used thermal analysis to define optimum piece part locations on circuit boards and optimum circuit-board locations within the unit for effective cooling.

Meinen, Carl. "Reliable Remote Monitoring and Control of Electrical Distribution." *Proceedings of the Annual Reliability and Maintainability Symposium,* 1980, pp. 448-452. Describes a broad view of derating including safety margins throughout the design as well as the capability to perform additional capabilities in the future. Includes examples.

Meyer, Stephen A. "Integrated Design Environment—Aircraft (IDEA): An Approach to Concurrent Engineering." Paper presented at the American Helicopter Society 46th Annual Forum and Technology Display, Washington, DC, May 21-23, 1990. Describes the integrated database used by diverse life-cycle team members who are participating early in the helicopter design.

Military Handbook 217. *Reliability Prediction of Electronic Equipment.* Describes techniques for obtaining base failure rates for various types of electronic components including the parts count method and the parts stress method.

Morris, Seymour F. *MIL-HDBK-217 Use and Application.* Rome Air Development Center Technical Brief, April 1990. Describes frequent misconceptions about the use and application of MIL-HDBK-217.

Naval Sea Systems Command. *Parts Application and Reliability Information Manual for Navy Electronic Equipment.* TE000-AB-GTP-010, September 1985. Describes derating techniques in which designers select parts and material to ensure the applied stress is less than rated for a specific application. Discusses benefits of derating, uses of derating curves, and possible trade-offs that may be necessary.

Nevins, James L. and Whitney, Daniel E., Eds. *Concurrent Design of Products and Processes.* New York: McGraw-Hill, 1989. Discusses multidisciplinary teams. Discusses why designers should consider producibility when choosing parts and materials. Discusses design for assembly practices and the benefits of manufacturing engineers participating in design decisions.

Owen, Jean V. "Assessing New Technologies." *Manufacturing Engineering,* June 1989, pp. 69-73. Discusses process problems with composites including how to handle them, form them, and automate their manufacture.

Oxford, David B. "Total Quality Management: Business Aspects and Implementation." *Integrated Circuit Screening Report,* Institute of Environmental Science, November 1988, pp. 2-1 to 2-8. Discusses the problems arising from antagonistic relations between suppliers and manufacturers and the advantages arising from cooperative alliances.

Phadke, Madhav S. *Quality Engineering Using Robust Design.* Englewood Cliffs, NJ: Prentice Hall, 1989. Describes the theoretical and practical aspects of quality engineering including robust design, matrix experiments using orthogonal arrays, and the design of dynamic systems.

Priest, John. *Engineering Design for Producibility and Reliability.* New York: Marcel Dekker, 1988. Discusses techniques for component management and control including environmental stress screening, group technology, and thermal analysis. Discusses principles and application of derating to ensure the stresses on the part are no greater than some percentage of the maximum rating. Discusses use of lists of preferred parts and materials that have met criteria for performance, reliability, timely delivery, and reasonable cost.

Raasch, Daniel. *The Defect Reduction Program.* Hopkins, MN: Honeywell, 1985. Describes the closed-loop defect reduction program at Honeywell.

RADC. *Parts Derating Guideline.* AFSC Pamphlet 800-27. Rome Air Development Center, 1983. Describes how to select derating levels for different parts families and applications using the part derating guide.

RADC. *Stress Screening of Electronic Hardware.* RADC Technical Report TR-82-27. Rome Air Development Center, 1982. Describes how to select levels for environmental stress screening of hardware.

Sanders, Robert T. and Green, Kent C. "Proper Packaging Enhances Productivity and Quality." *Material Handling,* August 1989, pp. 51-55. Discusses how damaging vibration may occur during product use, handling, or transporting. Discusses procedures for testing a product's susceptibility to vibration in which a product is subjected to gradually changing frequencies.

Sherry, W. M. and Hall, P. M. "Materials, Structures, and Mechanics of Solder Joints for Surface Mount Microelectronics." *Proceedings of the Third International Conference on Interconnection Technology in Electronics.* Fellbach, West Germany, February 18-20, 1986, pp. 47-81. Describes experiments that analyze how cycles to failure depend on thermal mismatches. Points out that as the use of surface mount devices increases, thermal mismatches and failures become likely at elevated temperatures.

Smoluk, George M. "Thermal Analysis: A New Key to Productivity." *Modern Plastics,* February 1989, pp. 67-73. Describes how thermal analysis is used to select materials for particular applications. Describes efforts to make thermal analysis systems more affordable and flexible and use personal computers for data analysis and storage.

Snead, Charles S. *Group Technology: Foundation for Competitive Manufacturing.* New York: Van Nostrand Reinhold, 1989. Describes group technology that codes and classifies parts into families. Discusses advantages for design and manufacturing engineers. Gives examples of group technology applications.

Testing to Verify Design and Manufacturing Readiness. Describes industry and government best practices for integrated test; test, analyze, and fix; failure reporting system; and uniform test report.

Therrien, Lois. "The Rival Japan Respects." *Business Week,* November 13, 1989, pp. 108-118. Discusses Motorola's six-sigma program to reduce its defect rate to 3.4 parts per million by 1992.

Wong, C. L. and Zimmerman, R. L. "Stress Screening Can Benefit a Pipeline Requirement." *Proceedings of the Annual Reliability and Maintainability Symposium,* 1987, pp. 125-129. Describes improvements in stress screening to reduce operational failures and repair time of the Phoenix missiles built by Hughes Aircraft.

Young, James F. and Shane, Robert S. Eds. *Materials and Processes, Part A: Materials.* 3rd ed. New York: Marcel Dekker, 1985. Discusses the information a designer needs to select, prove, and specify materials for product design.

Young, James F. and Shane, Robert S., Eds. *Materials and Processes, Part B: Processes.* 3rd ed. New York: Marcel Dekker, 1985. Discusses the processes involved in materials engineering and management. Focuses on information needed by people in product design, development, and production.

Subcontractor Control

Chapter

1

Introduction

My colleagues and I challenged 40 executive siminar participants to try and refute the following pointed statement: "Vertical integration is a failed strategy in today's environment. To survive, you must seriously and quickly consider subcontracting every task in the enterprise. It is virtually impossible to overdo subcontracting." To our surprise, banker, travel agent, brewer and retailer alike supported the notion.[1]

Program managers (PMs) of both civilian and Government contracts could also be included in Dr. Peter's list. As systems become more complex, they involve more and more technical specialization, and budgets shift the project climate.
This changing climate will see:

- More firms interested in subcontracting opportunities to build business bases

- More pressure by the customer to increase subcontracting by the prime contractor (or prime) to reduce cost and satisfy socioeconomic program objectives

- More firms subcontracting services traditionally performed in house

- More teaming agreements between primes and major subcontractors to bid and perform on major contracts

- More opportunities for primes to provide management and financial support to develop key subcontractors

---

[1]Tom Peters, "On Excellence," *San Jose Mercury News,* January 15, 1990.

- More subcontractor investment in product development to support major program opportunities.

The prime's relationship with the customer is that of an independent contractor, not as an agent of that customer. Only on rare occasions should there be a relationship between the customer and one of the prime's subcontractors.

- No contract exists between the customer and a subcontractor.[2]
- The prime is contractually obligated and compensated to manage the subcontractor(s).[2]
- Control is exercised by:
  Consent-to-contract approval
  Control of the prime.[2]

  The increased use of subcontractors provides the prime's management team with new challenges: project risk is increased if the subcontracts are not managed effectively and treated as the critical projects they are.

  Successful subcontract management has been described as the marriage between a prime and its subcontractor(s). In other words, the prime must be as committed to the success or profitability[3] of the subcontractor as to the profitability of the prime contract. This dual success is call a "win/win" situation.

  If the subcontractor is in a "lose" position so that no profit or gain is realized from the contract, the prime receives poor quality products and/or inadequate services. The prime must ensure that the effort to be performed and the contract under which the effort is performed involves the subcontractor(s) in a "win" situation.

What Is a Subcontractor?

A subcontractor is "an individual or business firm contracting to perform part or all of another's contract."[4] Although any effort farmed out to other companies is considered a subcontract, the types of subcontracts vary greatly according to the:

- Type of goods or services
- Technical effort required
- Risk associated with the subcontract.

---

[2]Center for Systems Management, *Subcontractor Management,* 1990.

[3]Profitability—Financial or technical knowledge gain.

[4]*Webster's New Collegiate Dictionary,* G. C. Merriam Co., 1971

Because of these variances, subcontractors fall into two basic categories, as follows.

- **Minor subcontractors or vendors**—Suppliers or distributors of commonly available goods and/or services, the requirements and specifications of which are well defined. These goods and/or services are usually off-the-shelf or commodity-type items obtained via fixed price contracts or purchase orders. Vendors provide commodity-type items. Examples: electronic parts, computers, software programs, and personnel services.

- **Major or critical subcontractors**—A contractor that provides some amount of research, development, and/or engineering services as well as providing goods and/or services. In this situation, all requirements and specifications may not be defined. Subcontractor services are purchased through a negotiated and "managed" contract which establishes various degrees of risk to be assumed by the subcontractor. The prime selects the contract type according to the degree of task difficulty.

- Similarly, large contractors often allow different divisions or separate locations of their company to perform work for their prime contract. These inter- and intracompany tasks usually provide financial or technical benefits to the prime. PMs must treat these inter- and intracompany dealings as *subcontracted* efforts. The PM must consider that the employees performing these tasks will probably not be under the PM's direction and may not be dedicated to meeting the PM's budget or cost goals.

Use this part for both vendors and major subcontractors. However, note that many items discussed in the Procedures chapter concentrate on areas predominantly concerning major subcontractors, as defined above. Tailor the process, as appropriate, to the specific procurement.

The best practices described in this book, while based on Government contracts, should be useful tools for commercial subcontracting to minimize risks.

Whether dealing with vendors or major subcontractors, a successful PM *understands* and *defines* the task to be performed and builds a contractual agreement that:

- Delivers the goods and/or services on schedule within budget
- Satisfies the technical requirements
- Minimizes the overall risk to the prime.

Primes subcontract effectively by using the proper tools and proven management techniques. This part gives PMs insight into those management techniques (refer to the Procedures chapter) and into lessons learned from other programs (refer to the Applications chapter). Application of these guidelines should result in better control of present and future subcontracts.

The Subcontract Management Team

The Subcontract Management Team (SCMT)—a subset of the Program Management Team (PMT)—initiates, plans, and executes the subcontracted task efforts. The activities of the SCMT are directed by a Subcontract Management Plan, just as the PMT activities are directed by a Program Management Plan. Figure 1 presents the organizational chart of a typical SCMT.

SCMT key members, as defined in this part, should include:

- Program Manager (PM)—Leader of the prime contract organization PMT

- Subcontracts Manager (SCM)—Member of the prime's PMT who is responsible for overseeing subcontract efforts; leader of the SCMT

- Subcontracts Administrator (SCA)/Buyer—SCMT member responsible for subcontract administration

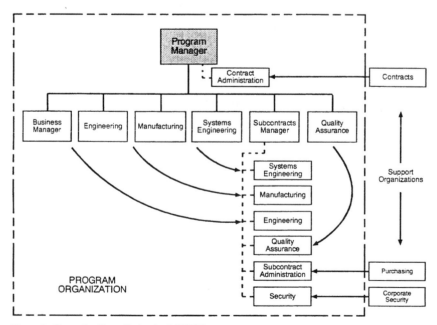

Figure 1 Organization of a typical SCMT.

- Business Manager—Member of the prime's staff responsible for both in-house and subcontracted business issues

- Technical Team Members—Members of the prime's staff responsible for overseeing the subcontracted areas, e.g., Quality, Manufacturing, Engineering, Integrated Logistics Support.

- On-site Representative(s)—Members of the prime SCMT located at the subcontractor's facility.

On-site Representatives should have full responsibility for the technical, cost, schedule, and quality aspects of the subcontract. Also, On-site Representatives should report directly to the prime's SCM.

For large subcontracts, significant in dollar volume and technical complexity, the SCM must use On-site Representative(s).

The SCMT has "cradle-to-grave" responsibility for the success or failure of a subcontracted effort. As shown in Figure 2, the prime's SCMT should be complemented by corresponding subcontractor personnel to coordinate execution of required subcontract functions.

The Risk of Subcontracting

Increased use of subcontractors increases risks because of:

- Loss of direct control over people performing the work

- Increased geographic dispersion

- Increased dependence for specialized, critical components

Prime Contractor Personnel	Subcontractor Personnel	Function
Program Manager	Senior Management	High-level Problem Resolution
Subcontract Manager	Program Manager	Subcontract Technical Direction
Subcontract Administrator	Contracts Administrator	Contractual Communication
Technical Team Members	Technical Performers	Information Exchange
Business Manager	Program Control	Business Management

Figure 2 Prime contractor/subcontractor organizational relationships.

- Poor flowdown of requirements by primes
- Poor prime control of subcontractors.[5]

Successful Subcontract Management

This part describes the functional operations necessary to perform successful subcontracts. Following these six steps will lead to successful subcontract management.

1. Define the effort to be subcontracted—Matching contract requirements to subcontractor capabilities through a structured mechanism.

2. Prepare the Request for Proposal (RFP)—Communicating those requirements to potential subcontractors.

3. Evaluate the proposal—Determining the potential subcontract's capability to satisfy the requirements as presented in the subcontractor's proposal.

4. Award the contract—Formally agreeing on the requirements to be satisfied and how and when they will be satisfied.

5. Control the subcontractor—Ensuring the requirements are met within cost, schedule, and performance parameters.

6. Conclude contract activities—Bringing closure to a mutually successful endeavor.

[5]Center for Systems Management, *Subcontractor Management*, 1990.

Procedures

Six chronological steps lead to effective subcontractor control.

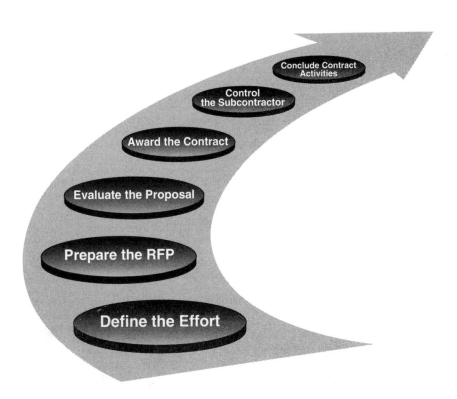

Step 1: Define the Effort to be Subcontracted

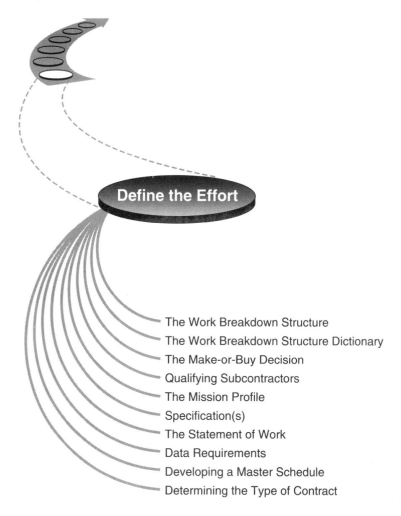

The Work Breakdown Structure
The Work Breakdown Structure Dictionary
The Make-or-Buy Decision
Qualifying Subcontractors
The Mission Profile
Specification(s)
The Statement of Work
Data Requirements
Developing a Master Schedule
Determining the Type of Contract

The work breakdown structure

As shown in Figure 3, the Work Breakdown Structure (WBS) is the central controlling framework for all aspects of program and subcontract management. The entire technical content of a development program is embodied in this structure.

The WBS includes all the work for a particular contract and relates all work elements to each other. It subdivides contract effort into:

- End item deliverables, for example:
  Hardware

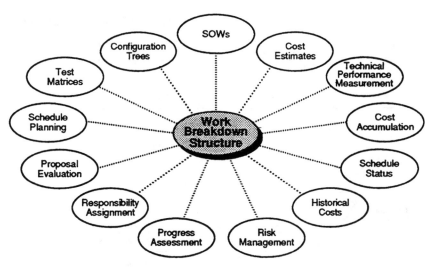

Figure 3 Controlling framework for all aspects of program management.

Software
Documentation
- Support, for example:
Training
Integrated Logistics Support
- Services, for example:
Systems Engineering
Program Management
- Facilities.

From a *customer* viewpoint, the WBS is a consistent and reliable means of obtaining adequate program visibility and of compiling historical costs for modeling and forecasting.

The *customer* usually establishes the summary levels of the WBS early in the program (as shown in Figure 4 for a sample aircraft development program). At this point, the *customer* has the greatest influence on the prime's extension to lower levels. This creates special considerations for subcontract management, as decisions to subcontract often affect many elements in the prime contract WBS.

Prime contract WBS. The prime should create extensions of the summary customer WBS which reflect how the prime will manage the work technically.

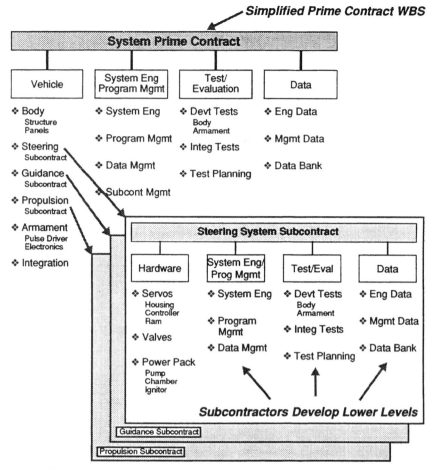

Figure 4 Sample WBS reflecting subcontract management approach.

This common framework:

- Manages requirements allocations, indentured end item lists, test requirements matrices, and configuration control
- Applies schedule, cost, and risk management to the technical elements selected for control
- Relates plans, tradeoffs, progress evaluation, and corrective action to the technical work.

The prime must identify all elements for a program in reasonable detail *before* they can be considered for requirements allocation or subcontracting.

What the work is must be considered before anything else.

The WBS is an ideal way to select work for subcontract; it also allows the prime flexibility in subsequent decisions whether or not to perform work in house.

The structure of the prime WBS significantly affects subcontract management; this must be considered during WBS negotiations. Figure 4 shows the generally preferred approach which creates a subsystem subcontract element at a high level in the prime contract WBS. Then the prime and subcontractor extend that element to suitable levels.

The resulting Subcontractor WBS (SWBS) provides good visibility and control. It truly is a usable tool for the subcontractor.

This approach to WBS/SWBS integration should be negotiated by the prime and the customer following prime contract and subcontract awards. It can be modified throughout the life of the program, depending on the management emphasis.

The SWBS reflects a natural management approach for the subcontractor, prime, and customer. Since costs accumulate at very low levels, accumulating program costs per the customer's original WBS can be easily satisfied through reallocation, when needed.

To accumulate costs by particular work elements (e.g., training, logistics), the customer may structure the WBS to disperse subcontract effort throughout the WBS. As illustrated in Figure 5, this dispersal through the WBS elements easily accumulates actual program costs but also forces use of an informal "management" SWBS for actual control purposes.

If the *customer* is unwilling to modify this type of WBS, the prime should make the customer aware of the potential for loss of visibility and incurrence of sizable administrative expense.

The work breakdown structure dictionary

While the WBS can be depicted as a tree of contract elements, the critical activity in developing the extended WBS is creating the WBS dictionary—a narrative description for each element of work, including references to the Statement of Work (SOW), specifications, data items, and responsible organizations.

To be complete, the WBS dictionary must include:

- What is required from others to start each task
- What exactly will be done in executing each task
- What will be produced and/or delivered to others
- A list of applicable specifications
- Who will receive the output.

Figure 6 shows a sample WBS dictionary element description sheet.

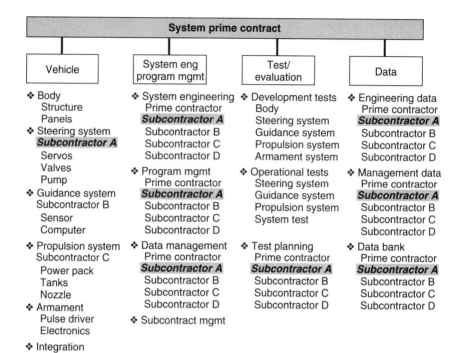

Figure 5 WBS for rollup of program costs—subcontract information.

Integration of the SWBS. In the SWBS, the subcontractor develops an extension of the prime's WBS in the same way the prime develops the WBS given by the customer. Since the SWBS is the controlling framework for the subcontractor, allow flexibility in the extension so it will be a useful tool, not a burden.

The make-or-buy decision

After program requirements are determined, the process of allocating those requirements to specific organizations begins. A part of this process is determining whether or not subcontractors will be used and, if so, defining their effort. This process begins with the Make-or-Buy decision.

The Make-or-Buy decision requires careful consideration because of the potential impact on program cost. Subcontracting results in:

- Reduced in-house business base and associated profits
- Reduced absorption of overhead costs (i.e., higher overhead)
- Loss of experience
- Potential competition for follow-on effort.

PROJECT: FXX CONTRACT NO.: ABC-91-444	WORK BREAKDOWN STRUCTURE DICTIONARY	DATE: MM/YY 10/91 SHEET 1 of _1_
WBS LEVEL: 2	**ELEMENT TITLE:** AIR VEHICLE	

ELEMENT DESCRIPTION:

AIR VEHICLE: The complete flyaway FXX for delivery to the US Government. The flyaway FXX constitutes the Avionics, Airframe, installed engines, and subsystems, as defined by the Detailed Specification for Model FXX Aircraft Weapons System, including all attendant addendums and the Avionic Specification.

WBS LEVEL:	**ASSOCIATED LOWER-LEVEL ELEMENTS (TITLE):**
3 3 3	AVIONICS AIRFRAME PROPULSION

Figure 6 Sample WBS dictionary element description sheet.

The near-term expediency vs. the long-term disadvantages to the company must be considered.

Usually a team participates in the Make-or-Buy decision. This team, lead by the PM, includes engineering, procurement, quality, manufacturing, business operations, and others.

While major items receive the first attention, review all items of significant cost, where both in-house and subcontractor capability exists, to determine the "best" choice for the program in light of the current business and cost objectives.

All items should be considered for subcontracting. The benefits gained may offset the associated risk.

The Make-or-Buy decision (see Figure 7) should consider cost and schedule, technical need, resources, and risk.

Table 1 presents some questions for consideration during the Make-or-Buy determination. These questions are not intended to lead to Make-or-Buy conclusions, but to present some important considerations for the Make-or-Buy team.

The Make-or-Buy process and the subsequent process of subcontractor selection is depicted in Figure 8.

Figure 7 Make-or-buy decision.

TABLE 1   Make-or-Buy Decision Questions

Considerations	Questions to be Answered
Technical Need	Are specialized skills or equipment required?
	Is the expertise and experience to perform the task available in house?
	Do any potential subcontractors have this expertise or experience?
	By subcontracting, would higher efficiency result through specialized skills or equipment?
Resources	Are the staffing and facilities to perform the task available in house?
	Would use of a subcontractor facilitate leveling of in house staffing requirements?
	Does the Prime Contract mandate use of subcontracts?
Cost and Schedule	Do subcontractors offer lower labor or overhead cost?
	Would use of a subcontractor have any effect on maintaining schedule?
	Are there small disadvantaged businesses that can perform this work?
Risk	Will risk increase through use of subcontracts?
	IF YES...
	Is an in house qualified subcontract management organization available to properly control the subcontractor?
	Will the geographic location of the subcontractor add to risk?
	Will use of a subcontractor for specialized skills result in increased dependency on a single source?
	Will supportability be affected?

Qualifying subcontractors

Once a buy decision has been made, potential subcontractors should be identified and qualified. Competition between qualified potential subcontractors usually reduces proposed costs and stimulates a variety of technical processes and solutions.

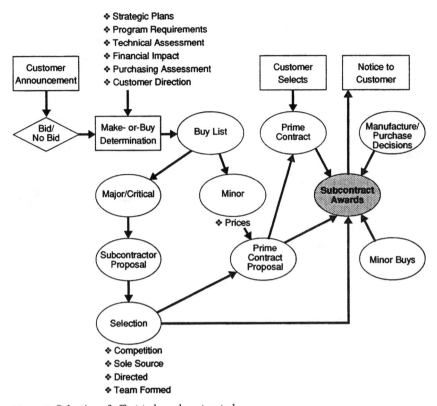

Figure 8 Selection of effort to be subcontracted.

Selecting the best subcontractor for any given task *begins* with properly selecting the bidding pool of potential subcontractors and *ends* with executing a contract with the winner of the competitive bidding.

Subcontractor data in a particular area of expertise is available from many sources. See Figure 9.

- Published documentation

  Trade magazines
  Department of Defense (DoD) Qualified Products Lists
  The Thomas Register
  Dun & Bradstreet reports

- The prime contractor's own corporate experience

  Purchase history records
  Quality rating reports
  Discussions with corporate personnel who have worked with subcontractors

Corporate Experience	Published Documentation

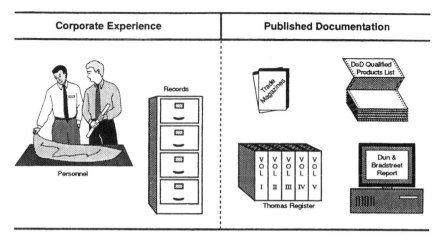

Figure 9 Sources of subcontractor information.

Corporate experience is an excellent source, *if the prime has had previous associations with a particular potential subcontractor.*

- Personal visits to potential subcontractors

  Depending on the nature, complexity, and size of the task to be subcontracted, a personal visit to a potential subcontractor may be in order.

- Contacts with subcontractor's representatives.

  A simple telephone contact with a potential subcontractor's representative usually results in information like brochures, pamphlets, etc.

In selecting potential subcontractors, the SCM or SCA evaluates such things as:

- Demonstrated past performance on similar projects

- Company capabilities

- The probability of the company successfully accomplishing the requirements.

Also, obtain evaluation information via supplier surveys and registration of facilities/supplier visitation reports. A sample supplier survey sheet is shown in Figure 10.

The type of evaluation needed to determine subcontractor capabilities is tailored to the nature, complexity, and size of the procurement. Depending on the complexity of the requirement(s) and the experience and confidence in the proposed subcontractor, send an evaluation team to perform an on-site review of subcontractor capability.

This process should result in the selection of the most qualified subcontractors to achieve project objectives at reasonable costs. Too often

Company name:	Location:
Small business:	Disadvantaged small business:
Facilities description:	Description of work force:
Sales:	Previous subcontracts:
Labor surplus:	
Financial status:	
Supplier visitation report:	SCA comments:
SCA's recommendation:	

Figure 10 Sample supplier survey sheet.

this important evaluation process does not get the attention and/or time to be done correctly.

Program risk will be reduced substantially through a careful identification and qualification of potential subcontractors.

The mission profile

System functional and environmental profiles drive the profiles and design requirements flowed down to subcontractors for a particular task.

Clear and concise design requirements are essential to obtain a realistic quote for a particular subcontracted task.

Refer to the *Design to Reduce Technical Risk* for the best practices to define system requirements effectively.

As shown in Figure 11 on page 252, the Mission Profile is initially defined at a high level (Step 1)[6]. As the functional analysis and risk analysis portion of system engineering are performed (Step 3), the Mission Profile is fine tuned to the real needs of the customer.

The prime should also follow this same process with each subcontractor to determine the true Mission Profile for each subcontract.

Specification(s)

Program-peculiar specifications should be developed and/or tailored based on the functional and environmental profiles and design requirements. They should cover only the effort to be subcontracted and

---

[6]AT&T, *Design to Reduce Technical Risk*, 1993.

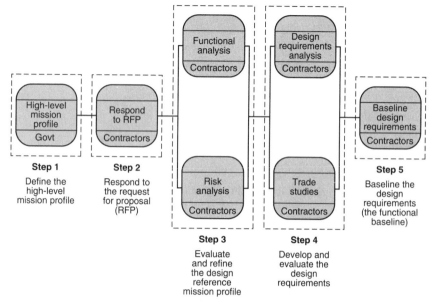

Figure 11 The design mission profile.

should include only the qualitative and quantitative design and performance requirements.

State performance parameters, acceptance criteria, and verification requirements clearly to avoid problems later in the program which could adversely affect cost and schedule. Figure 12 shows the types of specifications required for each phase of the development life cycle.

The statement of work

The SOW should follow the logical structure of the WBS and should establish and define all nonspecification requirements.

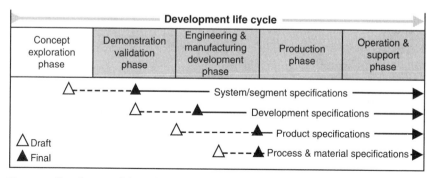

Figure 12 Development life-cycle specifications.

All tasks under an SOW paragraph should:

- Identify the work to be performed
- Be written in explicit terms
- Be expressed as minimal needs.

Depending on the contract, many tasks require adherence to military specifications, standards, etc., which are listed in the Reference Documents chapter of the SOW. Tailor any such requirements to meet the minimum needs of each individual SOW task.

Tailor specifications for subcontractor use. Flow down only applicable requirements.

Careful tailoring of the SOW:

- Promotes better understanding of the total task by the subcontractor
- Gives greater visibility into subcontractor cost and schedule
- Simplifies future cost adjustments brought about by program changes.

Data requirements

Requirements for data submittal must be defined in the Contract Data Requirements List (CDRL). Include all the information the subcontractor needs to furnish the data in the proper format.

- Sequence number used for tracking
- Data Item Description (DID)
- SOW reference
- Required submission dates
- Distribution and number of copies required
- Remarks

Data submittals are costly. Therefore, the prime should review data requirements thoroughly *before* passing them to the subcontractor. Tailoring DID requirements, for example, ensures that the prime subcontracts only for the actual needed items, and expensive "extras" are not included. Also, specify subcontractor format whenever possible.

Developing a master schedule

The subcontractor's Master Schedule should be a subset of the prime's Master Schedule. It should depict all program milestones and activities (i.e., documentation requirements, design reviews, fabrication, operational tests, deliveries). Subcontractor milestones and activities should be subordinated to prime milestones and activities, with completion

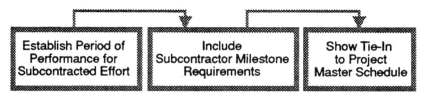

Figure 13 Requirements of the subcontractor master schedule.

dates that support corresponding prime due dates. *Prime / subcontractor dependencies should be established and incorporated into both the prime's and the subcontractor's network schedules.* See Figure 13.

Determining the type of contract

Developing the specifications, SOW, data requirements, and Master Schedule gives a detailed picture of the total task to be subcontracted and simplifies matching the type of contract to the total task.

There are two general types of contracts:

- Fixed Price contracts
- Cost Reimbursable contracts.

Fixed price contracts establish the work to be done and limit the total price paid for that work.

- The subcontractor *must complete* the work to receive payment.
- The pre-established price is not influenced by the subcontractor's costs or performance. If expenditures exceed the price, the subcontractor suffers a loss; if costs are minimized, subcontractor profits will increase.
- If the work is not satisfactorily completed on time, the subcontractor is contractually liable.

Cost reimbursable contracts are quite different.

- The subcontractor is paid for allowable costs incurred up to the established funding limit(s).
- Thereafter, the subcontractor is not obligated to complete the work unless additional funding is provided.
- The subcontractor usually does not incur any liabilities for defective or untimely work.

There are many variations of these two general types. The five most commonly used variations are shown in Figure 14 and briefly outlined in the following paragraphs.

	Fixed price	Cost reimbursement
	FFP	
Incentive contracts	FPI	CPIF
		CPFF
		CPAF

Figure 14 Contract types.

Fixed price contracts. A Firm Fixed Price (FFP) contract should be used when:

- The design baseline and specifications are firm.
- A fair price can be established.

An FFP contract requires very little administrative effort on the part of the prime contractor and transfers cost risk to the subcontractor. The subcontractor is contractually obligated to complete all specified work for the negotiated price, therefore *absorbing all cost risk*. The subcontractor must lower cost to maximize profit.

A Fixed Price Incentive (FPI) contract should be used when:

- FFP is not feasible because of an identified risk.
- Some risk sharing is desired.

The responsibility for meeting the target cost remains with the subcontractor; however, the prime is willing to share *some* of the risk. Additional costs incurred when the target cost is exceeded or savings accrued when it is underrun are shared, based on a predetermined formula. The amount of target cost, target fee, and ceiling price and a formula for sharing the cost over or under the target cost are negotiated as part of the contract. Cost, schedule, or performance may be incentivized. *Visibility is required by the prime contractor into both cost and schedule.*

Cost reimbursable contracts. A Cost Plus Incentive Fee (CPIF) contract transfers part of the cost risk to the subcontractor and should be used when:

- Development risks exist.
- Some risk sharing is desired.

Target cost and incentive fees are negotiated, as well as minimum and maximum fee points and share ratios. The negotiated share ratios between the prime and the subcontractor determine how risk of cost incurred above or below the target cost will be shared. All allowable costs will be paid; however, the subcontractor's fee is increased for an underrun and decreased for an overrun. This provides an incentive for the subcontractor to control costs. *Cost and schedule performance must be monitored carefully on a detailed level.*

The major differences between FPI and CPIF contracts are as follows.

- With an FPI-type contract, the contractor's obligation ends when the ceiling price is reached. As with the FFP-type contract, the subcontractor must complete the task with no additional funding, if costs exceed this ceiling price.

- With a CPIF-type contract, the contractor's obligation continues through payment of cost until the task is completed or cancelled. A CPIF-type contract has a minimum fee which the subcontractor receives regardless of the amount of overrun.

A Cost Plus Fixed Fee (CPFF) contract should be used only when the prime contract is cost reimbursable and when:

- Significant unknowns exist.

- The prime is willing to accept cost risk.

This contract allows for maximum flexibility and opportunity for change.

A CPFF contract invokes a negotiated fixed fee based on an agreed-to target cost. *However, the prime must manage the subcontractor closely, as there is very little incentive for the subcontractor to control costs.* Cost and schedule controls must be used, as all allowable costs must be paid. In other words, the prime assumes all cost risk.

A Cost Plus Award Fee (CPAF) contract *requires the contractor to assume all costs*; however, the subcontractor's fee will vary, based on the contractor's evaluation of the subcontractor's performance. Estimated cost, base, and maximum fee points are negotiated.

The fee is split into:

- A base fee (minimum fixed fee)

- An award fee (based on the prime's unilateral judgment evaluation of the subcontractor's performance).

The amount available for the award fee is the difference between the base and maximum fee levels. *The total amount of fee awarded will be based on the prime's periodic, subjective evaluation criteria.*

This type of contract is especially effective where the SOW and specification(s) do not contain a firm description of the work to be performed. Establish the criteria for evaluation, along with the award periods, *prior* to contract award.

The advantages of a CPAF-type contract are:

- Improved communication between the prime and the subcontractor through performance evaluation discussions
- Motivation of the subcontractor to furnish a higher quality product to receive an award fee which may be higher than the return from a CPFF-type contract.

The major disadvantage to a CPAF-type contract is the large administrative cost incurred by the prime for continual evaluation and processing of award information.

Choosing the contract type. When choosing the type of contract, consider the issues presented in Table 2.

TABLE 2 Considerations in Choosing Subcontract Type

	Subcontract Type			
Consideration	Fixed Price		Cost Reimbursable	
	FFP	FPI	CPIF	CPFF
Status of Design Baseline	No Changes Expected	Expect Minor Changes Only	Not Established	Not Established
Status of Specifications	No Changes Expected	Expect Minor Changes Only	Top-Level Only	Top-Level Only
Probability of Success	High	Good	Fair	Low
Complexity of Task	No New Development Required	Some Risk	High Risk	High Risk
Identification of Risk	Low	Medium	Medium	High
Ability to Estimate Cost	Excellent	Good	Poor	Poor
Type of Prime Contract	Any	FPI or CP	Cost Plus	CPFF
Expected Contract Changes	None	Some Expected	Many	Many
Cost and Schedule Required Visibility	Yes	Yes	Yes	Yes

Step 2: Prepare the Request for Proposal

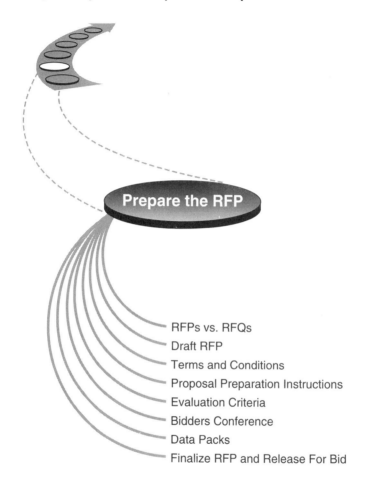

RFPs vs. RFQs
Draft RFP
Terms and Conditions
Proposal Preparation Instructions
Evaluation Criteria
Bidders Conference
Data Packs
Finalize RFP and Release For Bid

RFPs vs. RFQs

The SCM must determine the type of submittal required from potential subcontractors.

- Request for Proposal (RFP)—For written proposals or for open competition by major subcontractors
- Request for Quotation (RFQ)—For financial information only

RFQs can replace the longer and more detailed RFP in many instances, for example:

- Minor subcontracts or vendors

- Small Research and Development contracts
- Procurement of small quantities of high-tech hardware or software
- Noncompetitive subcontracts not requiring technical and/or management control.

The prime should consider whether RFQs are applicable to each subcontract and should tailor the requests, as needed. Noncompetitive areas where RFQs should be considered include:

- Sole Source Subcontract—A subcontract awarded to the only responsible source (contractor) capable of satisfying the prime's contractual requirements
- Directed Source Subcontract—A subcontract awarded by the prime which has been directed by the customer
- Team Partner—A subcontractor having a contractual agreement with the prime to assume some amount of risk.

As shown in Figure 15, the SCM directs the *development* of the RFP or RFQ. First, the SCM must decide on the RFP or RFQ format. Then inputs from other organizations are coordinated, along with developing the schedule, SOW, CDRLs, and other documentation.

Draft RFP

Often the customer may release a draft RFP, or the prime may develop a "strawman" which is used to refine the requirements and/or outline the tasks to be performed. A draft RFP is the foundation on which the actual RFP is built. The result is an accurate and realistic description of the desired work tasks. By providing an opportunity for potential subcontractors to review the requirements and to comment on whether those requirements can be achieved, the requirements and work tasks are refined and tailored. The draft RFP should include:

- Draft SOW
- Draft CDRLs
- Draft Proposal Preparation Instructions (PPIs)
- Draft Terms and Conditions (Ts&Cs)
- Master Schedule.

Terms and conditions

The prime contract's Ts&Cs restrict what may be done under the subcontract, as defined by applicable clauses in the public laws, Federal

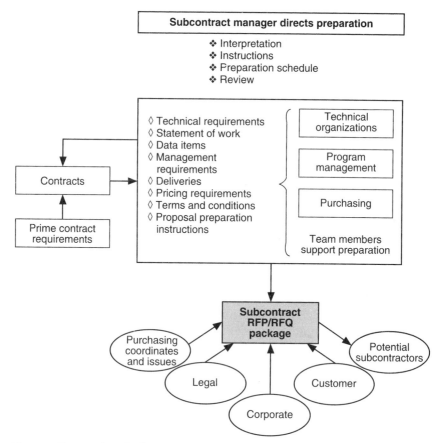

Figure 15 Preparation of subcontractor RFP/RFQ.

Acquisition Regulations (FARs), Defense Acquisition Regulations (DFARs), and other Government codes. The prime must evaluate and flow down these restrictions to subcontractors, as appropriate. See Figure 16.

Proposal preparation instructions

To ensure consistency between bidders' proposals and submittal of all pertinent information, the prime must prepare instructions for proposal preparation.

Different parts of a proposal require different information. As listed on page 261, these instructions or PPIs should include:

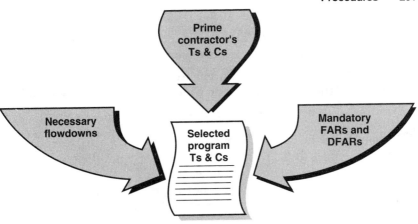

Figure 16 Subcontract terms and conditions.

TABLE 3 Subcontractor Cost and Pricing Requirements.

Cost/Price Element	Special Considerations	Cost/Price Element	Special Considerations
Materials	Recurring vs. nonrecurring	*Overhead rates*	When approved
Bill of materials used	Long-lead items		How approved
Source of prices	Attrition quantities		
Quantity rationale		*Royalties*	
Labor	Task/phase	*Other Direct Costs*	Assumptions
Source of rates	Line item/Lot/SOW	Travel	Quantity
Direct labor quantity	Year/month/quarter	Computer	Duration
Skill mix	Fiscal/calendar	Packaging	
Estimating methods	Basis of estimate	Consultant	
	Standards	Reprographics	
	Learning	*Fee Proposed*	Percent Incentives/sharing Award fee

- Cost Volume—The prime should instruct the bidders on how to present cost and price data for easy evaluation for reasonableness. Cost estimates (see Table 3) must be broken down by individual task or task element and must include both the estimating methods and rates. Unless directed by the PPIs, the cost volume probably will not contain the information necessary to perform a cost comparison or audit.

- Technical Volume—The prime should instruct the bidders on format and page count for the technical volume.

- Management Volume—The prime should instruct the bidders on format and page count for the management volume.
- General Grading Criteria—To ensure that bidders emphasize the right areas, the PPIs should state which areas carry the most weight.
- Examples—To depict the required proposal format.

Without PPIs the bidders' proposals will not have consistent content and format and will be difficult to evaluate. By developing PPIs, the proposals can easily be audited and evaluated.

The SCM involves key members of the proposal evaluation team in writing PPIs. Concurrent development of the PPIs and proposal evaluation criteria:

- Ensures that all information necessary for evaluation is included in the proposal
- Helps emphasize important technical areas.

Table 4 illustrates how the PPIs and evaluation standards work together to provide the needed information for proposal analysis.
The PPIs should include:

- Administrative directions—number of copies, page limitations, type size, delivery dates and places, etc.
- Detailed instructions describing specific information to be provided. For example, "The Offeror shall describe the process to be utilized in satisfying the Cost/Schedule Control System Criteria requirements."
- Specific instructions for the Cost Volume to satisfy audit requirements.

Because the information required for an audit can be proprietary, the subcontractor may need to prepare a sanitized version of the cost proposal for the prime, as well as a complete version for the customer's

TABLE 4   How PPIs and Evaluation Standards Work Together

PPI	Evaluation Standard
The Offeror shall describe the process to be used in satisfying the Cost/Schedule Control System Criteria (C/SCSC).	The standard is met when the Offeror describes a C/SCSC system with current Government validation or the steps to be implemented to receive validation within 90 days of contract award.
The Offeror shall describe the program organization structure and the lines of authority and responsibility.	The standard is met when the Offeror describes a clear path of authority and responsibility to and from the PM.

auditors. The prime, in turn, must safeguard the technical and financial information entrusted to him by his potential subcontractors.

Evaluation criteria

Prior to RFP release, the SCM directs the SCMT to establish a scoring system to set an objective method for awarding a subcontract. When developing these evaluation criteria, determine the method for scoring:

- Comparative rating—A selection method in which all proposals are compared to each other, and the best proposal is selected.

- Absolute evaluation—A selection method in which all subcontractors are scored against an evaluation standard, and the proposal that best meets all the requirements is selected.

Also, review the RFP to ensure that the areas to be evaluated are actually included. This scoring system should include:

- Objective ratings for RFP requirements

- Guidelines for technical compliance such as design/development/fabrication approach, risk areas, meeting specifications

- Specifics that will be evaluated

- Other criteria such as experience, location, and past performance.

If the prime's proposal evaluation team does not generate a rating system, the team will lose sight of the technical/cost/management needs. As a result, the contract will probably be awarded to the subcontractor with the lowest price, but not necessarily the best technical solution.

By using evaluation criteria, the evaluation team will score according to how the requirements are met, not on cost alone. Result: contract award will more likely go to the most qualified subcontractor.

Establish the evaluation criteria before release of the RFP, and include the criteria in the RFP. The RFP should state the general importance of the areas being evaluated, but not the scoring standards. Typically, list the areas in order of importance, e.g., Technical, Management, Manufacturing, and Cost.

Bidders conference

The SCA should hold a Bidders Conference to clarify the RFP. This conference presents the opportunity to tailor or eliminate specifications or requirements questioned by potential subcontractors.

The Bidders Conference should include a review of the subcontract requirements listed in Table 5.

TABLE 5   Subcontract Requirements

Technical Requirements:
Reliability and Maintainability
Supportability
Performance
Environmental
Management Requirements:
Cost Control
Schedule
Material Control
Quality Assurance
Configuration Management
Data Requirements
CDRLs
Electronic Media
Drawings
Contract Ts&Cs and FAR Restrictions:
Proprietary Information
Corporate Flowdowns
Nondisclosure
Terms and Conditions
Mandatory FARs/DFARs
Patent Rights

Data packs

A Data Pack contains the information referenced in the RFP, including:

- Specifications
- Drawings
- Computer programs
- Electronic media or data
- Compliance documentation
- Other documentation.

When the RFP is released, the Data Pack *must* contain complete and accurate information. Take special care with specifications and drawings (including revision numbers).

Failure to release quality Data Packs results in unnecessary subcontractor questions and proposals that do not fulfill the true needs of the subcontracted effort.

The SCM should control the development, release, and revision of Data Pack information by applying the methods discussed in *Design to Reduce Technical Risk.*

Finalize RFP and release for bid

After the Bidders Conference, finalize the RFP for release to potential bidders. Carefully review the finalized RFP to ensure that the changes clarifying the tasks and enhancing the performance of the subcontract have been included.

The completed RFP should include the information shown in Figure 17.

The SCM must ensure that the RFP is complete and adequate and must balance the subcontracting process with the support of the SCMT and the PM. To preserve the integrity of the bidding and the competitive process, if applicable, direct all requests from any bidder for information about the RFP to the SCA.

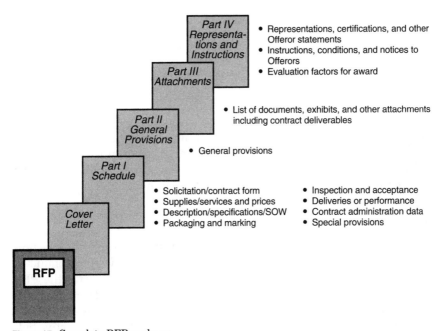

Figure 17 Complete RFP package.

Step 3: Evaluate the Proposal

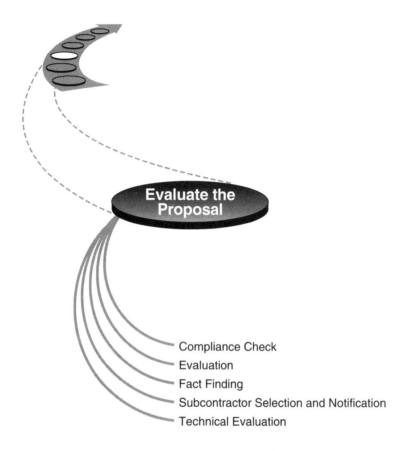

Compliance Check
Evaluation
Fact Finding
Subcontractor Selection and Notification
Technical Evaluation

Figure 18 shows the roles that different organizations may play in the proposal/quote evaluation process. These are typical functions for evaluating a large proposal; however, these roles may vary and should be tailored for each subcontractor as directed by the SCM.

Compliance check

As soon as the proposals are received, the SCM directs a subset of the proposal evaluation team (usually the SCA and purchasing) to conduct a "quick and dirty" initial evaluation to ensure compliance with the PPI. This team checks such things as page limits and requested certifications; this team is *not* responsible for *evaluating* the responses. Noncompliant proposals are subject to being rejected and returned to the respective bidders. See Figure 19 for a list of sample compliance criteria.

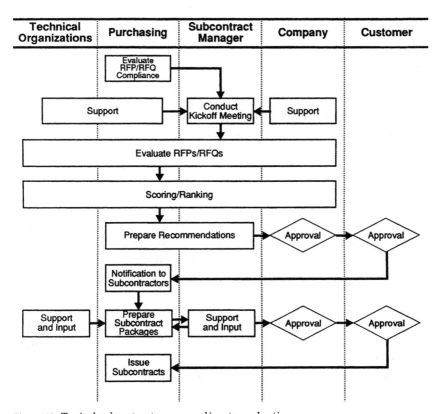

Figure 18 Typical subcontractor proposal/quote evaluation process.

Proposal Compliance Checklist		
	Yes	No
1. Proposal received on time?	X	
2. Contains all volumes?	X	
3. Within page limit?	X	
4. $ values removed from tech volumes?	X	
5. Includes required clearances?	X	
6. Correct Print/Typeset?	X	
7. Foldouts within limits?	X	
8. Compliance with other specific guidance which was included in the PPIs?	X	

Figure 19 Sample initial proposal evaluation criteria.

Evaluation

Proposal evaluation is based on:

- The type of subcontractor (minor or major)
- The complexity of the subcontract.

After passing the initial check, the proposal is evaluated by a team selected by the SCM. The proposal is graded according to the evaluation criteria discussed in Step 2.

The composition of the evaluation team should reflect the subcontractor type and the complexity of work involved, as follows.

- Minor subcontractors—Evaluations of minor subcontractor or vendor proposals can be performed by the SCA or members of his staff. This evaluation should be an absolute evaluation against finalized requirements and specifications. Evaluation of vendors or suppliers are usually much simpler and less time consuming than major subcontractors because the strictly defined requirements usually represent commodity or off-the-shelf items.

- Major subcontractors—Evaluations of major subcontractor proposals are usually performed by a project team headed by the SCM which includes the SCA, Business Manager, auditors, and technical team members. These evaluations involve fact finding, quality audits, and detailed comparisons to the evaluation criteria, sometimes over a long period of time.

Usually the PM has final approval authority for proposal acceptance or rejection. However, in some companies, the evaluation team may act only as a recommendation board; the actual decision maker selects the subcontract winner based on the evaluation team's inputs.

Technical, management, and cost considerations. Proposals are evaluated according to the technical, management, and cost considerations listed in Table 6. These factors should be considered when scoring a proposal.

Audits. To verify financial information and technical capabilities, perform financial and quality audits prior to contract award.

Financial audits help determine if the subcontractor has proposed reasonable material costs, labor rates, overhead, and general and administrative costs. In many cases subcontractors will not allow a prime to conduct financial audits. In such cases, the customer would perform the audit, with the prime assisting, as necessary.

Also, perform quality audits to determine if the subcontractor has the capability to perform the task described in the SOW.

TABLE 6   Technical, Management, and Cost Evaluation Considerations

Technical Evaluation	Management Evaluation	Cost Evaluation
Risk Assessment	Management Structure	Labor Category
Soundness	Schedule Management	Hours
Complexity of Approach	Cost/Funds Management	Materials
Compliance with Specifications	Subcontract Management	Services
Technical Staff	Material Management	Overhead
Concurrent Experience	Automated Information Systems	Cost of Money
Past Experience		Subcontracted Efforts
Testing and Design Verification Methodology		Rates
Quality		Bases of Estimates
Systems Engineering		Reasonableness of Costs
Configuration Management		
Integrated Logistics Support		

Fact finding

For complete proposal evaluation, all questions arising during the evaluation must be clarified. The SCA gathers the necessary facts to answer these questions either by letter or, more likely, during a face-to-face meeting with the potential subcontractor. The following steps should be taken to complete a fact-finding mission.

1. Assemble a fact-finding team. To ensure proper evaluation, the prime's fact-finding team should consist of expertise in the following areas:

   Technical

   Financial

   Management

   Subcontracts management.

2. Prepare for the face-to-face meeting.

   The fact-finding team must be familiar with each subcontractor *before* an on-site visit. Each team member should:

   Read the specifics of the proposal

   Conduct preliminary technical/management/cost evaluation

   Prepare an agenda and briefing material.

3. Prepare questions.

   Prepare questions on any area of the proposal not fully understood, and submit them to the subcontractor through the subcontract manager prior to the fact-finding meeting. Proposal areas usually requiring clarification include:

Personnel (labor category and number of hours)

Facilities

Schedule

Technical

Costs/Price.

4. Hold the meeting.

The SCM should serve as the fact-finding team leader and conduct the meeting.

The fact-finding team leader should be the single point of contact between the fact-finding team and the subcontractor.

Subcontractor selection and notification

When the fact findings, audits, and proposal reviews are completed, the prime should hold an internal *formal* subcontractor selection award meeting. During this selection meeting, discuss and document the justifications and criteria for awarding the subcontract. This meeting should be chaired by the PM and should include all staff members who participated in writing the RFP and evaluating the submitted proposals.

Technical Evaluation

I. Overview
    A. Basic summary of proposal
    B. Summary of costs and prices
    C. Period of performance
II. Evaluation Criteria
    A. List of ground rules
    B. Evaluation ground rules
    C. Estimate of resources
III. Findings
    A. Proposal contents
    B. Technical approach discussion
    C. Analysis of task by task effort
IV. Recommendations
    A. Changes in effort
    B. Justifications for change
V. Summary
    A. Costs summary
    B. Change summary

Figure 20 Contents of a technical evaluation.

Promptly notify all subcontract proposal competitors of the selection. The subcontract award winner should also be briefed on probable contract negotiations and a definitization schedule.

The prime must be prepared to debrief each unsuccessful bidder, pointing out strengths and weaknesses of unsuccessful proposals, if selection was made for reasons other than cost. The SCM or SCA should handle and document all debriefing communications with bidders, both written and oral.

Technical evaluation

The SCA must direct the technical staff to document the justification for selecting a particular subcontractor in a Technical Evaluation. This document is often audited by the customer and may include the information presented in Figure 20.

Step 4: Award the Contract

Award the Contract

Consider Letter Subcontracts
Technical/Management Negotiations
Cost/Price Negotiations
Terms and Conditions Negotiations
Subcontract Award

After subcontractor evaluation and selection, contract execution can begin either via a fully executed contract or, when time is of the essence, via a letter contract.

Consider letter subcontracts

At times the PM determines that a letter contract is appropriate to award the subcontract. Use a letter contract when:

- Immediate contract execution is required to meet schedules.
- Adequate funding is available.
- Specific beginning tasks are defined.
- An Interim Period of Performance is defined.
- Limitation of expenditures is established.

The letter contract enables the subcontractor to perform tasks per the SOW during negotiation and/or definitization of a full-blown contract. During the execution period of letter contracts, costs are reimbursed, but no fee is paid until the fully executed contract has been definitized.

Negotiations

Prior to negotiations, the PM should select a negotiating team to:

- Assign roles
- Develop strategy
- Establish targets
- Agree on process issues.

This team should consist of the SCM, the SCA, and key staff members.
As shown in Figure 21, negotiating a contract, or definitizing a letter contract, requires:

- Agreement on the effort involved
- Agreement on the price of that effort
- Agreement on any contractual restrictions on the subcontractor.

Technical/management negotiations. Normally, the SCM leads Technical/ Management Negotiations, with guidance from the SCA and technical staff. Agreement must be reached on the level of resources required to accomplish the subcontracted task satisfactorily. During negotiations, the prime should review:

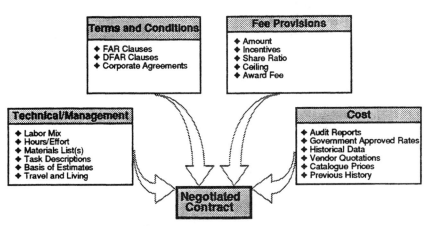

Figure 21 Inputs to negotiated contracts.

- The proposal
- The past history of the subcontractor
- Industry standards
- An estimate of the level of resources required if the task were accomplished in house.

Cost and price negotiations. The SCA and his staff usually conduct Cost and Pricing Negotiations. Audits conducted by the prime, or assist audits conducted by the customer, form the basis for these discussions. (Assist audits are required when a subcontractor is not willing to release proprietary cost information to the prime.)

Terms and conditions negotiations. Terms and Conditions Negotiations are also conducted by the SCA, with assistance from the SCM. The legal and patent organizations also provide assistance, as needed.

Subcontract award

Once the full contract or letter contract has undergone Technical and Management Negotiations, Cost and Price Negotiations, and Terms and Conditions Negotiations, it is ready for final approval and signature. When signed by representative officers of the respective parties, the contract is a legal document upon which all parties mutually agreed.

After signature, the contract becomes active for the contract period of performance. Work can now be performed in accordance with the contract, and costs and fees are now billable.

Step 5: Control the Subcontractor

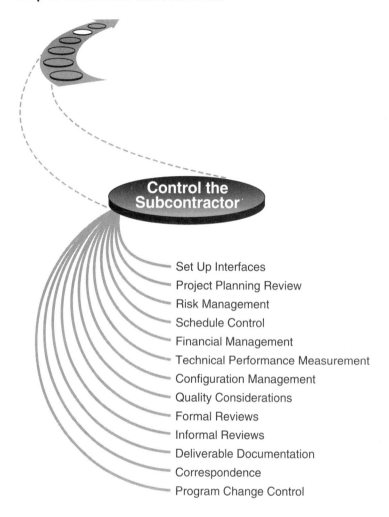

Set Up Interfaces
Project Planning Review
Risk Management
Schedule Control
Financial Management
Technical Performance Measurement
Configuration Management
Quality Considerations
Formal Reviews
Informal Reviews
Deliverable Documentation
Correspondence
Program Change Control

The subcontract management control process is a continuous reporting and approval network. Figure 22 shows the interface relationship between the subcontractor and the SCM in this control process.

Set up interfaces

Ensure that the subcontractor appoints a PM with authority to speak for and make commitments on behalf of the subcontractor. Also, name an individual to speak for and make commitments on behalf of the prime (normally the SCM).

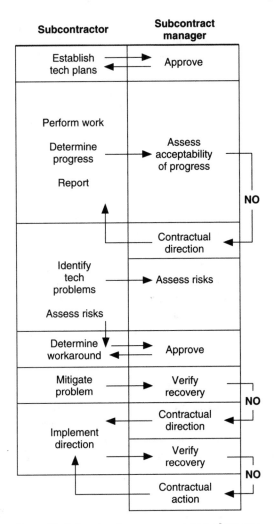

Figure 22 Subcontract management control process.

Other prime contractor organizations or individuals should not be allowed to give direction or change what the SCM has established with the subcontractor.

Single points of contact for the prime and each subcontractor are essential to promote clear understanding.

Under the SCM and subcontractor's PM, set up additional working points of contact between similar organizations; i.e., engineering to engineering, quality to quality.

It is extremely important for the SCM to maintain open communication in all areas. However, any *significant changes* must be handled between the SCM and the subcontractor's PM and *must be formalized through contract channels.*

Communication between functional organizations is encouraged; however, these interfaces must be limited to the exchange of information only.

An important function of the SCM and the SCMT is establishing processes for requirements control to prevent out-of-scope effort. Fundamentally, this is a function of *clear and specific* SOWs, specifications, and WBS dictionaries for both the prime and each subcontractor.

To assess the impact of out-of-scope decisions on the total program, the SCM should always get approval of the prime's PM on any such changes. If a subcontract out-of-scope decision translates into the prime being out-of-scope with the customer, defer the decision until the *customer* approves the change.

The SCMT must learn and understand the subcontractor's management structure to be able to effectively interface with it. However, the SCMT should not attempt to mold the subcontractor to the team's own image. If the subcontractor's management methods prove to be inadequate, the prime should be prepared to provide guidance, direction, or other assistance, as needed.

Project planning review

Hold a project planning review meeting as soon as possible after the subcontract is awarded. A suggested agenda is presented in Table 7.

The importance of this meeting cannot be over emphasized. It establishes the standards of performance expected during the entire subcontract period.

TABLE 7   Suggested Project Review Meeting Agenda Items

Subcontractor:	Project plan—how the SOW will be implemented, in detail
	Review of all specifications, drawings, etc.
	Detailed schedules for implementation (especially the first 6 months)
	A staffing profile of the subcontractor's task (including key personnel)
	A time phased budget
	Management of configuration control, quality, reliability, safety, data management, cost control, etc.
Prime:	Project management structure, methods, and who provides direction
	Overview of the prime contract and the subcontractor's role

Hold the review meeting at the subcontractor's location and include:

- The subcontractor's management team and project leaders
- The prime's SCM
- Other key members of the prime contractor's organization.

For a major/critical subcontract, this meeting should also be attended by key members of the prime's and subcontractor's senior management. The prime contractor should emphasize:

- *Success*—The prime wants the subcontractor to be successful and to make a profit.
- *Teamwork*—Working together brings success to all parties.
- *Visibility*—The prime needs to know about problems to help solve them. Discuss methods of monitoring technical, schedule, and cost performance.
- *Open communication*—Interfaces should be established on all levels. Discuss formal and informal meetings and who should participate.
- *Quality*—Establish quality standards and expectations.
- *Assistance*—Make the prime's technical and management resources available to the subcontractor, as needed.

At the conclusion of the meeting, the prime and subcontractor management teams should be in agreement on all aspects of contract management and implementation.

Risk management

Risk management lessens the probability that an undesirable event will occur and/or mitigates its effects. By using risk management techniques, primes and subcontractors can identify potential problems *before* they occur, establish technical alternatives, and estimate probable impacts on cost and schedule (see Figure 23).[7]

The prime should tailor and flow down the prime's Risk Management Plan to each subcontractor.

Risk planning.   Each subcontractor should develop a Risk Management Plan and Risk Organization to monitor and handle identified risk ele-

---

[7]Defense Systems Management College, *Risk Management Concepts and Guidance,* 1989.

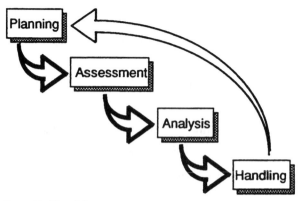

Figure 23 The risk management process.

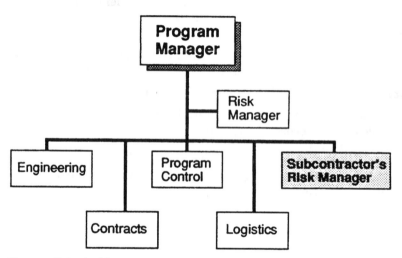

Figure 24 Prime's risk management.

ments. The subcontractor's Risk Manager should have access to and may sit on the prime's Risk Management Board (see Figure 24).

Risk assessment. The subcontractor must identify potential risk areas and provide that information to the prime. These risk areas include technical, operational, supportability, cost, and schedule facets, as shown in Figure 25. Each facet affects the performance of the opposite facets.[8]

---

[8]Defense Systems Management College, *Risk Management Concepts and Guidance,* 1989.

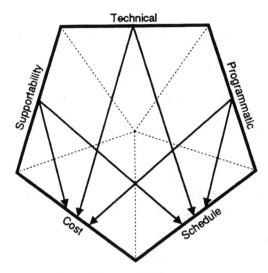

Figure 25 Potential areas of risk.

Classify risks. Subcontractors should have a rating system, tailored to the specific program, to classify each risk element High, Medium, or Low in regards to the probability of occurrence and severity of consequence.

Risk analysis. The subcontractor must analyze each risk area to determine possible mitigation techniques. Contractors can use techniques such as network analysis, modeling, and trade studies to determine possible fall-backs or alternatives.

The subcontractor must have detailed plans for risk mitigation and decision points for implementing these plans. The subcontractor must also have methods for tracking the risks using cost, schedule, and/or technical progress.

The subcontractor should use some objective method for tracking technical progress against predetermined technical goals [Technical Performance Measurement(TPM)]. This progress should also be the basis for schedule and earned-value status. If required by the subcontract, technical, schedule, and earned-value items should form the basis for reporting and monitoring by the subcontractor's Schedule and Cost Control System (SCCS).

Schedule control

Define and evaluate the subcontractor's schedule and scheduling process in the proposal phase, agree to these processes during the Program Planning Review, and monitor them throughout the performance of the program. Similarly, include requirements for regular statusing and forecasting in the subcontract, with the prime retaining approval authority for all subcontractor schedules.

Subcontractor milestones and activities should be subordinated to prime milestones and activities, with subcontractor completion dates established to support their corresponding prime due dates. Establish prime/subcontractor dependencies, and incorporate them into both the prime's and the subcontractor's network schedules.

On a continuing basis, review schedule status and contingency plan development and timely implementation. While the prime should be aware of the implementation of any contingency, actions affecting the prime's schedule should require *prior* approval.

The subcontractor's schedule system should include:

- Master Program Schedule
- Intermediate Network Schedule
- Detailed schedules.

Master program schedule. The subcontractor's Master Program Schedule should be a subset of the prime's Master Program Schedule and should depict all program milestones and activities (e.g., documentation requirements, design reviews, fabrication, operational tests, deliveries, etc.).

Intermediate network schedule. The subcontractor should develop a Network Schedule supporting its Master Program Schedule. This Network Schedule must be subordinate to the prime's and must indicate *internal* dependencies as well as dependencies to the prime network.

Detailed schedules. The detailed or working-level schedules are developed and maintained at the work-package level and are subordinate to the Network Schedule.

Financial management

All program problems (schedule, technical, etc.) will eventually be reflected in cost.

To adequately control program costs, both the prime and subcontractor must implement a financial management system. For major subcontracts, the system may be based on the C/SCSC, or for smaller subcontracts, the system may be subcontractor-unique.

In either case, *do not limit financial management merely to cost reporting.* Rather, financial management should include organizing, planning, budgeting, recording and reporting, analyzing, and revising.

This process begins with RFP preparation and concludes only after contract closeout.

During the RFP and proposal phases, the prime reviews the contract requirements and its own financial management system and determines the financial management requirements to be flowed down to each subcontractor. The RFP should specify such financial management concerns as level of reporting and types of control.

The subcontractor's plan for financial management should be presented in the proposal and should be evaluated by the prime. In addition to evaluating the proposed plan and processes, evaluate the cost proposal for the structure and use of the WBS in quoting; resource levels, allocation, and timing; and reasonableness based on the Basis of Estimates.

Upon issuance of a subcontract, the subcontractor must establish a technical, schedule, and cost baseline and implement the processes necessary to maintain that baseline. At the Program Planning Review, the prime should review:

- Budget realism

- Establishing benchmarks to monitor technical progress

- Establishing and controlling management reserve

- Timing of resource allocation

- The work authorization process

- Accountability

- Reporting processes and timing

- Variance analysis and corrective action.

After approving the subcontractor's plans and processes, focus on ensuring financial management by the subcontractor. To do this, monitor key performance indicators such as:

- Missed milestones

- Earned value criteria

- Test results

- Financial performance.

Variance thresholds, established in the subcontract, are triggers for highlighting and controlling problems through appropriate corrective action, either initiated by the subcontractor or directed by the prime.

Where necessary, corrective action by the prime may include:

- Fee

- Progress payment

- Award fee adjustments
- Termination, in extreme cases.

Financial forecasting and funds management.   Financial forecasting and funds management are aspects of subcontractor control that are critical to the prime's financial operations. Even so, they are often poorly managed. For a prime subcontracting significant amounts of work, faulty projections can produce substantial monetary errors. Financial forecasting and funds management consist of:

- Predicting the flow of company (and customer) funds required to pay or subcontract effort
- Integrating these predictions into company financial projections to determine rates at which profit will be recorded ultimately.

While this definition is stated rather simply, accomplishing these objectives can become extremely complicated given:

- Numerous subcontracts to manage
- Incentivized subcontracts (particularly with intricate incentive structures)
- Multiple programs with large amounts of subcontracted effort
- Subcontracts experiencing performance problems.

Frequently, all these situations exist simultaneously.

This management predicament can be solved with careful program analysis by the SCMT to develop forecasts of:

- Timephased base subcontractor expenditures
- Specific scheduled points at which to determine liquidation and incentive costs, including any award fees
- Subcontractor billings for expenditures, liquidation, and fee(s)
- Anticipated payment of these costs.

All payments should be done in conjunction with the prime's financial and accounting staff. Any payments affect the forecasts of the prime's profit rates as influenced by the spending of outside procurement dollars and, therefore, must be carefully monitored.

Financial forecasting is not a simple process. However, a competent SCMT, working closely with the subcontractor and effectively using all available information, will be able to develop very accurate predictions as subcontractor work progresses. At the program level, these predictions are then used to determine very accurate predictions of customer funding needed to sustain the prime contract effort.

Technical performance measurement

TPM is a tool that both the prime and subcontractor can use to track technical accomplishment vs. funding and schedule milestones. The prime must establish goals defining the level of technical achievement per the subcontract WBS. These goals or subcontractor technical milestones are usually called technical parameters. Figure 26 shows the technical parameters for a sample shipboard fire control system.

First, the prime establishes high-level technical parameters which are flowed down to the subcontractor. Then the subcontractor must develop intermediate- and lower-level technical parameters early in the subcontract.

By using technical parameters, the prime can track whether or not a particular subcontractor is meeting schedule milestones. The prime can also track the direct relationship between funding expenditures and the amount of technical progress for each parameter. Figure 27 shows a sample milestone schedule slip and cost overrun based on a missed technical parameter milestone.

Some typical areas with which technical parameters are correlated include:

- Design Reviews

- Breadboard analysis

- Hardware validations

- Performance tests

- Environmental tests

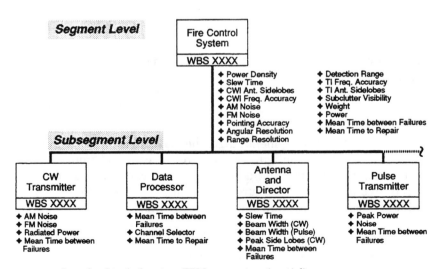

Figure 26 Sample of typical system TPM parameters (partial).

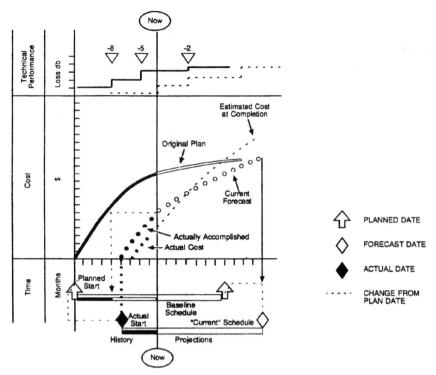

Figure 27 Sample milestone schedule slip.

- Qualification tests
- Reliability tests
- Computer software tests
- Integrated Logistics Support testing and demonstrations
- Operational readiness tests.[9]

Configuration management

Depending on the development stage, the subcontractor must have an accurate method for configuration control or configuration management (CM). The prime must determine which configuration controls are required and then make sure those controls are in place.

All contractors (prime and subcontractors) must:

---

[9]Defense Systems Management College, *Systems Engineering Management Guide,* 1986.

- *Identify and document* the functional and physical characteristics of a configuration item
- *Control changes* to those characteristics
- *Record and report change* processing and implementation status
- *Conduct configuration audits* to verify that the product conforms to the approved documentation
- *Monitor and audit subcontractor* CM programs.

Involve CM early in the design phase to reduce the risk of extensive changes later in the program. Figure 28 gives an overview of the elements of CM.[10] The prime should also ensure that subcontractors invoke CM techniques in their design program.

Quality considerations

To control and ensure quality product design, the prime should consider:

- Flowing down configuration requirements
- Placing a subcontractor representative on the prime's Change Control Board

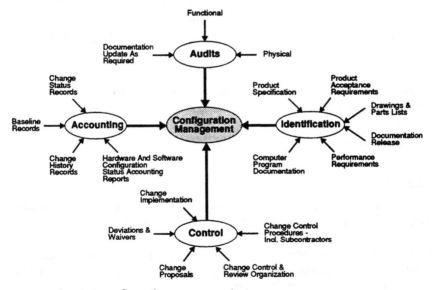

Figure 28 Inputs to configuration management.

---

[10]AT&T, *Design to Reduce Technical Risk*, 1993.

TABLE 8   Key for Receiving Subcontractor Quality Products

Source inspection of all parts and materials before shipment
Component-level, subassembly-level, and/or assembly-level integration testing before delivery
Addressing reliability, maintainability, and supportability issues during the design phase, not after production begins or delivery
Acceptance testing of parts and/or materials upon receipt from subcontractor
Accurate and complete documentation (drawings, specification, correct revision, etc.)
On-site Representatives at each subcontractor when problems arise or full time residents, as required by the contract

- Using status accounting systems for tracking development of configuration items.

The prime must flow down the customer's basic quality requirements to the subcontractor via a tailored Quality Assurance Program. Table 8 presents some key elements of such a program.
The subcontractor's quality effort should include:

- *A quality plan* —Guidelines for purchased material, vendor control, nonconforming materials, corrective actions, drawings and drawings changes, equipment calibration, electrostatic discharge, assembly test, inspection and handling, and marking equipment

- *A quality organization* —The organization that will control the subcontractor's quality effort, including responsibilities and duties for each member

- *Quality audits* —Methods for monitoring and controlling the manufacturing processes

- *Methods for traceability and tracking* —The documentation process for required customer traceability.

Formal reviews

The prime must also control the design program through open communication. One form of communication is formal reviews.

Project reviews. Schedule project reviews on a periodic basis (monthly, semi-monthly, quarterly—depending on the size and complexity of the task), and include a standard agenda.

- Limit project reviews to presentation and discussion of key management indicators demonstrating progress on the program.

- Refer items outside the contractually agreed upon schedules and budgets to the appropriate SCMT member for investigation and reporting to the SCM.

- Request and monitor recovery plans from the subcontractor.

- Avoid discussion of technical issues to keep attention focused on schedule; cost; and status of corrective action, recovery, or risk management plans.

The subcontractor's PM should conduct project reviews. Participants include:

- Prime's SCM and SCMT members
- Prime's technical leaders
- Prime's purchasing representative(s)
- Customer representatives
- Subcontractor's key people.

Alternate project reviews between the prime and subcontractor locations, if possible, so that personnel at each location can attend at least every other meeting.

Technical reviews.   Technical reviews should be scheduled according to SOW requirements. They include, but are not limited to, design reviews, test readiness reviews, system requirements reviews, acceptance test reviews. (Refer to *Design to Reduce Technical Risk* for detailed information on all reviews from the initial stages of design to final production.)

Schedule technical reviews in conjunction with significant milestones, if possible, and *prior* to the prime's corresponding reviews with the customer. They should occur at the end of each step in the design process before a commitment is made to proceed to the next step.

Figure 29[11] shows an example of technical reviews for a *development program*. A *production program* would have a different set of review requirements.

The purpose of technical reviews is to:

- Determine design maturity
- Ensure the design is technically adequate

---

[11]Defense Systems Management College, *Systems Engineering Management Guide*, 1986.

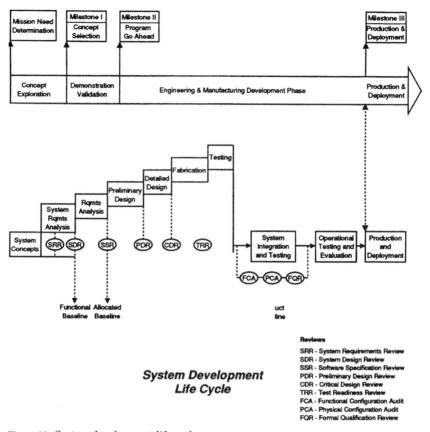

**System Development
Life Cycle**

Reviews
SRR - System Requirements Review
SDR - System Design Review
SSR - Software Specification Review
PDR - Preliminary Design Review
CDR - Critical Design Review
TRR - Test Readiness Review
FCA - Functional Configuration Audit
PCA - Physical Configuration Audit
FQR - Formal Qualification Review

Figure 29 System development life cycle.

- Ensure the design will meet customer requirements for performance, quality, cost, and availability.[12]

Therefore, only technically competent personnel should be selected as reviewers.

To ensure accurate information transfer, the prime should also include key subcontractor technical personnel in the prime's technical reviews.

---

[12]AT&T, *Design Reference Guide, Design Series,* Transition from Development to Production, 1989.

Informal reviews

Also promote communication and control through informal reviews. Informal reviews supplement the formal project review system. They occur between formal review sessions to:

- Exchange timely data on technical progress, costs, and schedule
- Resolve minor issues and items not requiring formal review
- Promote a close working relationship between the prime's SCM and the subcontractor's PM.

Technical reviews. Hold informal technical reviews to investigate and resolve technical issues. They are the ideal mechanism for the prime to learn what is really happening within the subcontractor's design organization and what the problem areas are.

Informal technical reviews are usually technical interchange meetings which:

- Take place at any time during the contract period
- Promote much technical discussion rather than information presentation
- Include prime technical expertise for input into decisions without taking over the design or the task
- Generate minutes for traceability, documenting the issues discussed, questions raised, agreements made, and action items assigned.

Any agreements reached must be within the present SOW and baseline. If problems arise which can only be solved by changing the baseline, these problems and the suggested solutions must be taken to the prime's SCM and the subcontractor's PM for resolution.

Changes to the baseline cannot be made at informal review meetings.

Deliverable documentation

The prime should set up a documentation control system to handle all deliverable documentation received from the subcontractor. This system should be under the control of the SCA, who makes distribution and tracks all actions required by the data deliverables.

Much of the data is furnished in draft format and needs to be reviewed, with the resulting comments returned to the subcontractor. The subcontractor cannot finalize this data until receiving the review comments; therefore, *any delay can negatively impact the subcontractor's schedule.*

The SCM should coordinate all review comments and forward them to the subcontractor in a timely manner.

Deliverables furnished in final form, not requiring review, comment, and return to the subcontractor, should be reviewed by the prime's functional groups.

If acceptable, the SCM should be notified that the data meets the specified requirements. If unacceptable because requirements are not met, the SCM should reject and return the deliverable along with reason(s) for the rejection.

Correspondence

Channel official program direction correspondence between the prime and subcontractor through a single point of contact. The prime's SCM and the subcontractor's PM should be the official channel for program direction.

Information exchange between functional organizations, however, does not have to pass through this channel, as this type of communication does not constitute program direction.

Transmit all official contract correspondence (e.g., directives, notifications, modifications, and requests) through the designated contractual focal points:

- The prime's SCA
- The subcontractor's Contract Administrator.

Program change control

Change happens on any major subcontract. The amount of change depends on how well the original subcontract was written and how well the contracted task is progressing.

The prime's SCM is responsible for controlling all changes impacting subcontractor performance as defined in the original subcontract requirements. This includes contract changes requiring a modification to the contract price, as well as engineering changes with no cost impact.

Basically, subcontract program changes fall into two categories:

- Those that are out-of-scope for both the subcontractor and the prime
- Those that are out-of-scope to the subcontractor but in-scope to the prime.

Figure 30 and Figure 31 depict the steps necessary to effectively incorporate subcontract changes.

Remember: All program changes must be authorized by the SCM and processed through the SCA to the subcontractor's Contract Administrator.

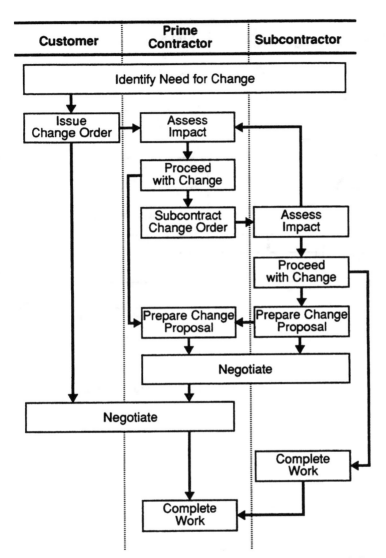

Figure 30 Incorporation of changes that affect the prime contract and the subcontract.

Customer	Prime Contractor	Subcontractor

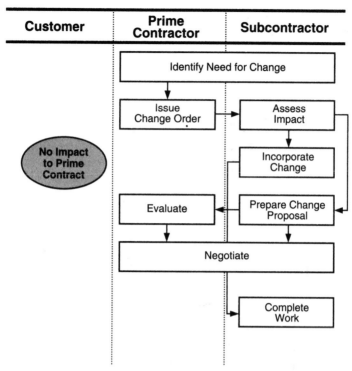

Figure 31 Incorporation of changes that result in no change to the prime contract.

Step 6: Conclude Contract Activities

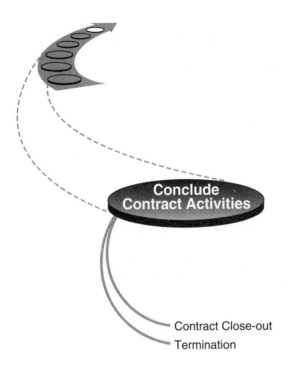

Contract Close-out
Termination

Contract close-out

After a subcontractor has physically and financially completed all tasks:

- Fully document the contract.
- Carefully consider disposition of customer and prime property and classified material.
- If the subcontractor wants to retain classified material, forward their request to the customer's contracting officer and advise the subcontractor of the decision.

Requirements for close-out vary, depending upon the type of subcontract. Normally, the prime's purchasing organization performs these activities.

Termination

For both fixed-price and cost-reimbursement contracts, termination of a contract by the prime falls into:

- Termination for convenience
- Termination for default.

Termination for convenience. This type of termination allows:

- Unilateral termination by the prime when:
  Supplies are no longer needed
  It is in the company's interest to do so
- A complete termination of all items on contract which have not been completed and accepted
- A partial termination of some of the items that have not been completed and accepted
- Termination of a subcontract immediately after notification from the customer that the prime contract is terminated in whole or in part (any delay can result in additional cost which may not be recoverable).

Termination for default. This type of termination allows the prime to terminate a subcontract when the subcontractor has failed to:

- Deliver the required supplies
- Perform the required services
- Otherwise perform the agreement.

Termination for default should be coordinated with the Legal department to ensure adequacy of documentation.

Before terminating a contract for default, however, the subcontractor must be given an opportunity to correct the performance deficiency. The prime should:

- Notify the subcontractor by letter of the possibility of such action
- Tell the subcontractor how much time is available to cure the problem
- Advise the subcontractor of contractual liabilities
- Request an immediate explanation of the failure to perform satisfactorily.

Chapter

3

Application

To give PMs guidance and insight into "lessons learned" on actual subcontracts, this chapter presents a selection of case studies dealing with subcontractor control and a typical subcontract procurement flow diagram.

Three case studies show the effects of subcontract management practices, both good and bad, on the success of a prime contract. Although these case studies are based on actual situations, some program details have been modified for anonymity.

These case studies pinpoint the effects of subcontract management practices on the overall program. *However, remember that the prime's approach to subcontract management is a logical extension of the company's approach to management in general.*

Figure 32 (on pages 304 and 305) presents a subcontract management flow diagram which shows the steps used successfully in large major/critical subcontracts. Although not always used, this diagram includes a draft RFP cycle which defines contract requirements more clearly and prevents costly misunderstandings and subcontract changes.

Good Examples of Bad Subcontractor Control

The following two case studies show the consequences of little or no subcontractor control and the improper use of subcontractor control, respectively.

Case Study #1

*Taking over the subcontractor's
management responsibilities spells
problems for an electronics surveillance
system program.*

A Government prime contract for a major electronics surveillance and detection system, to be installed, tested, and used on Air Force reconnaissance aircraft, was issued to a prime experienced in integration. In turn, the prime awarded a major subcontract for control processor hardware, the heart of the system. This subcontract comprised nearly 35 percent of the total contracted effort.

The problem

The prime was required to use a formal SCCS—IAW DODI 7000.2. These requirements were flowed down to the subcontractor, but the prime strongly directed the subcontractor "not to waste a lot of time on that." Instead, the prime would generate most of the schedule and cost information based on informal subcontractor inputs.

Basic initial flaws in this contractor /subcontractor relationship:

- An inadequate subcontractor SCCS:

    Only limited review by the prime of the subcontractor's SCCS

    Subcontractor unwilling to correct major deficiencies found by the prime's review ("Do you want hardware, or do you want paper?" was the subcontractor's standard response).

- Subcontractor actually driving the contract control:

    Subcontractor technically superior to the prime

    Subcontractor one of a very few sources of the type of processor subcontracted for

    Prime conceded to perform necessary schedule and cost planning and progress assessment themselves rather than to irritate the subcontractor and prime's management by insisting on adhering to the subcontract requirements.

The consequences

The subcontractor began to experience technical problems, then schedule delays, and finally obvious cost overruns.

To promote false economy and to save money, the prime reduced contract performance information submittal requirements ("C-Spec" requirements).

- The subcontractor was then able to reduce the size of its program control staff substantially and save a lot of money.

- After 3 years the subcontractor submitted only actual cost data; program status information was reported only through program status reviews.

Again to save money, the prime approved a subcontractor request to suspend all formal cost/schedule management activities and reporting. (The prime directed the SCM to "just get the job done!")

- The prime now developed all schedules for the subcontracted effort.

- Schedule status was obtained informally by telephone.

- Invoices were automatically approved and forwarded to accounting for payment.

- When schedule dates were missed, new dates were given.

After $4^1/_2$ years of virtually no subcontractor control:

- The subcontract was nearly 80 percent overrun with no significant hardware delivered because of major technical problems.

- The prime contract was 60 percent overrun and over 20 months behind schedule. (Because of so many slips, it was almost impossible to determine what schedule to measure against.)

- Most importantly, *the customer had not received any significant hardware.*

The ultimate results

Nearly 6 years into the contract, those system components that were delivered, installed, and tested did not perform to the specifications of the prime contract. The processors did not work as specified, experienced continual failures, and caused repeated downtime of test aircraft.

The customer began to doubt the real status of the contract's development program and took over control.

- Customer fact finding and cost teams began to investigate the prime and then the subcontractor. *Conclusion:* progress against specific plans could not be determined.

The subcontractor had no planning baseline.

The only apparent cost controls were head-count limitations imposed on the subcontractor by the prime.

- Both the prime and the subcontractor were directed to reinstate technical and schedule planning baselines and then develop cost estimates to complete the work.

- Simultaneously, the customer formulated its own schedule and cost estimates.

- The customer finally determined that the technical requirements of the contract could not be successfully fulfilled within a reasonable time or within the funding available.

The final conclusion: After 8 years, the prime contract was concluded, nearly 3 years behind schedule and more than 130 percent overrun. The technical performance was so unsatisfactory that the follow-on production was disapproved by the Defense Acquisition Board. *The program was considered a colossal failure.*

Case Study #2

Overzealous flowdowns of specifications and deliverable data requirements bog down a major spacecraft program.

A major multiservice spacecraft program was structured by the Government to direct nearly 70 percent of the prime contract effort to subcontractors.

The problem

Because of the program's size, complexity, performance environment, and interrelationships of the end users, the prime contract included extensive Government specification, standards, and data item requirements. However, this created more problems than it solved.

- Spacecraft performance specifications in the prime contract were often ambiguous ("To be Determined") and sometimes contradictory.

- To counter recent Government accusations that the prime practiced weak subcontract management on other existing contracts, the prime was overly anxious to show their spacecraft customer (and their subcontractors) that all subcontractors would be managed very strongly.

- The prime chose to flow down subcontract requirements that exactly duplicated the requirements in the prime contract, *even to the point of actually duplicating pages and including them in the subcontract.*

- The result was tremendous confusion regarding:
  Work scope
  Compliance requirements
  Component specifications
  Delivery destinations for data items.

The prime would not consider any subcontractor objections, and the customer was apparently pleased at the prime's firm resolve. Standard replies to subcontractor objections became, "It's what the customer wants," "We're not asking you, we're telling you," We're paying you to figure that out," and "Don't ask again."

The consequences

As the program progressed, several nightmares began to emerge.

- Customer requirements became better defined.
  Some performance specifications were understated.
  Others were absurdly rigorous.
- To speed up specification changes, revisions were verbally flowed down to subcontractors.
- As the design matured and hardware was being produced and tested, several problems surfaced.
  Several subcontractors just could not meet some of the stringent specifications.
  Integration testing experienced substantial delays.
- Even though the subcontractors could show that the spacecraft would perform satisfactorily with reduced requirements, the prime refused to relax the original requirements.
- Despite all efforts, the subcontractor for a critical electronic component could not meet timing specifications. The prime would allow no deviation, even though he could no longer determine the origin of those specifications.
- One subcontract was terminated because the subcontractor stopped trying to improve his performance.
- Submittal of required data items just didn't measure up.
  Many were marginally compliant.
  Many were continually late or months overdue.
  Some were never submitted and probably never would be.

All this contributed to eroded schedules and overrun costs. Then, to save money, the prime directed the subcontractors to reduce headcounts, *but no corresponding reductions in work scope or delivery schedules were made.*

The ultimate results

Because the prime still refused to consider realistic requirements, overruns grew, further subcontractor headcount reductions were mandated, and schedules fell further behind. By this time, both the prime and the subcontractors had very small work forces performing only those tasks absolutely necessary to complete, test, deliver, and integrate flight hardware.

By default, the prime began easing subcontract requirements.

- To save time, the prime began accepting reduced performance if overall system performance was not adversely affected.

- Because of extensive hand reworking during integration testing, many components resembled custom-made laboratory models rather than configured end items.

- System configuration was largely "as-built" rather than "as-designed."

- Documentation was badly out-of-date.

The customer finally imposed major program funding reductions. Both the prime and the subcontractors were directed to develop "could cost" strategies to identify and eliminate excess effort beyond what was needed to fulfill the most basic contract requirements. As a result, a considerable amount of work was eliminated.

The final conclusion: The prime contract underwent a major restructuring so that the subcontracts were explicitly tailored to reflect the actual work needed. While this *finally* resulted in a cooperative and productive prime/subcontractor team effort, the overall program suffered a 2-year schedule slip and a 100-percent overrun. *However, the customer determined that the originally planned large production program was economically unfeasible and cancelled it.*

A Good Example of Good Subcontractor Control

The following case study shows the benefits of well-planned and well-administered subcontractor control.

Case Study #3

A hands-on subcontract management
approach results in a successful aircraft
production program.

A major aircraft production program consisted of a prime contract and over 50 major/critical subcontractors who performed nearly 60 percent of the work.

- Nearly all subcontracts were FFP, although the wing and auxiliary power unit contracts were CPIF.

- The prime contract was FPI with an 80/20 share ratio.

The prime contract represented a substantial company outlay of capital investments, as well as a significant percentage of the company's business base.

The need and the management solution

All levels of the prime's management strongly emphasized the need to effectively control subcontractor performance, and a dedicated subcontract organization was formed to manage all aspects of subcontracting.

The prime developed and administered a formal, consistent, and documented approach to managing this enormous effort.

Each subcontract had its own SCMT, consisting of the following, at a minimum. (A published list of team members was kept current.)

- SCM

- SCA

- Price/cost analyst

- Scheduler

- Program control administrator.

The implementation

From the start clear requirements, schedules, and reporting mechanisms were set in motion to guide and control this effort.

- All subcontracts included identical requirements for measuring technical and schedule progress and for ensuring that expenditures were consistent with the work performed.

- The non-FFP contracts included requirements for formal management control systems subject to review, approval, and periodic on-site reviews by the prime.

- The prime contract WBS, consisting of individual elements for each subcontract, easily showed the relationships between subcontractor performance elements.

- Subcontractor performance results were presented to the prime's executive management weekly and to the customer during frequent status reviews.

- All subcontracts included provisions for progress payments, as determined by the SCM for acceptable technical and schedule progress. This helped eliminate large cash-flow surprises on this fixed-price, tightly scheduled program.

- To verify technical progress, all subcontractors had to provide detailed milestone schedules showing significant activities leading to end-item delivery, for example:

  Tool design/fabrication

  Long-lead procurement

  Detail fabrication

  Assembly

  Integration

  Test.

- Major structure subcontractors presented this same information in Line-of-Balance format, for integration into the prime Line-of-Balance schedules.

- All subcontractor schedules were approval data items and were reviewed by both the SCMT and the prime's master scheduling group.

- Once approved, baseline plans could not be changed without prior SCM concurrence.

- Depending on subcontract size and criticality, each subcontractor submitted statused schedules weekly or monthly, with telephone updates between submittals.

- The prime verified reported subcontractor progress.

  For large subcontracts, On-site Representatives verified reported progress.

  For small subcontracts, the SCM verified progress during on-site visits.

- All subcontractors provided expenditure information.

  Actual costs to date

  Forecasts of costs to be incurred

  Forecasts of progress and liquidation billings

If actual expenditures differed substantially from forecasts (10 percent), progress payments were withheld pending acceptable rationale and resubmittal of revised expenditure forecasts.

The ultimate results

Many factors favored successful subcontractor management of this prime contract.

- High-level commitment from all levels of management

- Clear definition of subcontract management responsibilities

- A management approach formally adopted and consistently applied

- Flowdown of requirements to subcontractors which provided ample subcontract visibility but not excessive demands

- As much on-site presence as necessary to confirm acceptable work performance

- Judicious use of strong (fiscal) controls.

The result: The prime contract was able to meet or exceed every key program schedule event. First flight was 4 months ahead of schedule. This continued through delivery with the last unit 3 months ahead of schedule. *All work was completed within 3 percent of the target cost, resulting in a very happy customer and a very profitable program for the prime.*

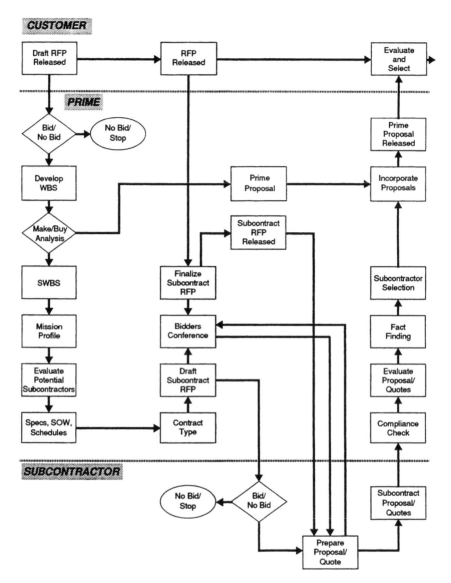

Figure 32 Subcontract management flow diagram.

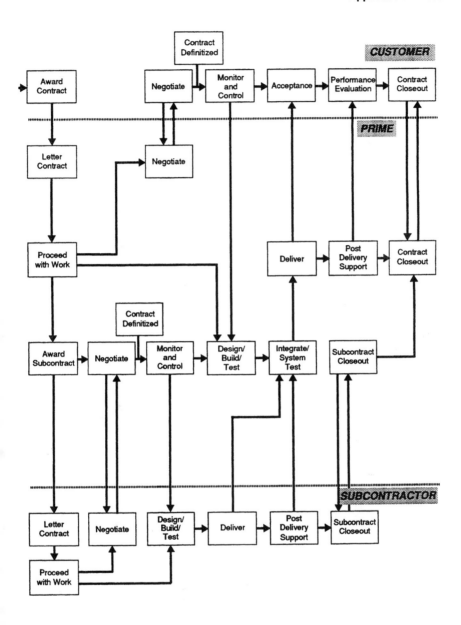

4

Summary

In our high-technology age, management of subcontractors is becoming more important every day. Very few companies have the capabilities to design, test, and manufacture a high-tech product/system while remaining competitive in the world marketplace without subcontracting part of the effort.

Effective subcontractor control ensures that subcontractors produce goods or services that meet the prime's specified requirements, within schedule and cost limitations and with minimal risk to the prime.

In other words, the prime can satisfy its customers only if it is able to subcontract effectively.

Follow Best Practices

The procedures discussed in this part describe the best practices for subcontractor control.

Most importantly, remember that prior to beginning any subcontracting effort, the prime must first establish and commit to:

- A Subcontract Management organization
- Roles of the Subcontract Management organization
- Program and Subcontract Management Plans.

Summary of Procedures

The steps presented in the Procedures chapter are summarized below. These summarizations include:

- Each step broken down into individual tasks
- The recommended personnel for directing and performing each task
- Key points for each task.

TABLE 9   Summary of Subcontractor Control Procedures

Task	Directed by/ Performed by	Key Points
Step 1: Define the Effort		
The WBS	PM/SCM	The controlling framework for all aspects of program management.
		Reflects the way work is technically managed.
WBS Dictionary	PM /SCM	Narrative description of each element in the WBS.
Make-or-Buy	PM/MOB Team	Consider all elements of the WBS.
		Consider technical need, cost and schedule, resources, and risk.
Qualification of Subcontractors	SCM/SCA	Can substantially reduce program risk.
Mission Profile	PM/Staff	Define system requirements effectively.
Specifications	PM/Staff	Tailor to the effort to be subcontracted.
Statement of Work	PM/SCMT	Follow the structure of the WBS.
		Identify all work.
		Tailor for minimal needs.
Data Requirements	PM/SCMT	Data is costly.
		Tailor for minimal needs.
Master Schedule	PM/SCMT	Depict milestones and activities.
Type of Contract	PM/Staff	Match contract type to task.
		Consider the status of the development program.
Step 2: Prepare the Request for Proposal		
RFPs/RFQs	SCM	Minor or major subcontractor.
		Sole or directed source.
Draft RFP	SCM/SCMT	Refines requirements and outlines the tasks.
Terms and Conditions	SCA/Legal Dept.	FARS and DFARS.
		Company Ts&Cs.
PPIs	SCM/SCMT	Defines guidelines for submittal of cost, technical, and management volumes.
Evaluation Criteria	SCM/SCMT	Establish objective criteria.
		Establish technical compliance guidelines.
Bidders Conference	SCA	Clarifies RFP to potential bidders.
Data Packs	SCM/SCMT	Complete and accurate specifications, drawings, computer programs, etc.
Finalize RFP	SCM/SCMT	Incorporate suggested changes by potential subcontractors.

TABLE 9 Summary of Subcontractor Control Procedures (*Continued*)

Task	Directed by/ Performed by	Key Points
	Step 3: Evaluate the Proposal	
Compliance Check	SCM/SCA	Noncompliant proposals subject to rejection.
Evaluation	SCM/SCMT	Based on type of subcontract and complexity of tasks. Evaluations conducted on technical, management, and cost.
Fact Finding	SCA/Part of SCMT	Variety of expertise. Prepare early and develop questions.
Subcontractor Selection and Notification	SCM or SCA/SCMT	Document all results. Notify and debrief unsuccessful bidders.
Technical Evaluation	SCA/Technical Staff	Justification of subcontractor proposal.
	Step 4: Award the Contract	
Letter Contract	PM/SCA	Use when contract execution is needed to meet schedules.
Technical/Management Negotiations	SCM/Staff	Agree on resource levels prior to negotiations.
Cost Negotiations	SCA/Staff	Audits form the basis for these negotiations.
Contract Execution	SCA/PM	Negotiations completed.
	Step 5: Control the Subcontractor	
Set up Interfaces	SCM	Single points of contact for program direction.
Project Planning Review	SCM	Establishes standards of performance expected.
Risk Management	SCM/SCMT	Lessons probability of undesirable events.
Schedule Control	SCM/SCMT	Interdependencies of prime and subcontractor networks. Depicts all program milestones and activities.
Financial Management	SCM/SCMT	Monitor key performance indicators. Take corrective action.
TPM	SCM/SCMT	Tracks technical accomplishment vs. funding and schedule.
Configuration Management	Technical Staff	Identifies and documents configuration.
Quality	Technical Staff	Specify sufficient inspection at all levels.

TABLE 9   Summary of Subcontractor Control Procedures (*Continued*)

Task	Directed by/ Performed by	Key Points
	Step 5: Control the Subcontractor	
Formal Reviews	SCM	Periodic Status Reviews.
		Technical Reviews IAW SOW.
Informal Reviews	SCMT/Technical Team	Exchange timely data on technical progress, costs, and schedule.
		No changes to baseline.
Deliverables	SCM/SCMT	Set up documentation control system.
Correspondence	SCM/SCA	Use single point of contact.
Program Change Control	SCM/SCA	Authorized by SCM only.
		Processed by SCA to subcontractor's Contract Administrator.
	Step 6: Conclude Contract Activities	
Contract Close-out	SCA/Purchasing	Fully document close-out.
Termination	PM and SCM/SCA	Termination for convenience.
	Legal Department	Termination for default.

Chapter

5

References

Center for Systems Management. *Subcontractor Management*. Santa Clara, CA: 1990. Discusses the methods for properly performing subcontract management.

Defense Systems Management College. *Risk Management Concepts and Guidance*. Ft. Belvoir, VA: U.S. Government Printing Office, 1989. Discusses the tools and processes used in the Government approach to risk management.

Defense Systems Management College. *Systems Engineering Management Guide*. Ft. Belvoir, VA: U.S. Government Printing Office, 1986. Discusses the tools and processes used in the Government approach to systems engineering.

Department of the Navy. *Best Practices: How to Avoid Surprises in the World's Most Complicated Technical Process*. (NAVSO P-6071), March 1986. Discusses how to avoid traps and risks by implementing best practices for 47 areas or templates, including topics in design, test, production, logistics, facilities, and management.

Design to Reduce Technical Risks. Discusses the processes and tools required to define mission profile, design requirements, and trade studies and to perform configuration management.

Kelley, C.G. and Sammet, G., Jr. *Dos and Don'ts in Subcontract Management*. New York, NY: AMACOM, 1980. Discusses the methods for properly performing subcontract management and the associated pitfalls in subcontracting.

Peters, T. "On Excellence," *San Jose Mercury News*. San Jose, CA: January 15, 1990.

Webster's Seventh New Collegiate Dictionary. Springfield, MA: Merriam Company, 1971.

.

6

Glossary

Acronyms and Abbreviations

CDR   Critical Design Review
CDRL   Contract Data Requirements List
CM   Configuration Management
CPAF   Cost Plus Award Fee
CPFF   Cost Plus Fixed Fee
CPIF   Cost Plus Incentive Fee
C/SCSC   Cost/Schedule Control System Criteria
DFAR   Defense Acquisition Regulation
DID   Data Item Description
DoD,DOD   Department of Defense
FAR   Federal Acquisition Regulation
FCA   Functional Configuration Audit
FFP   Firm Fixed Price
FPI   Fixed Price Incentive
FQR   Formal Qualifications Review
MIL-STD   Military Standard
MOB   Make-or-Buy
PCA   Physical Configuration Audit
PDR   Preliminary Design Review
PM   Program Manager
PMT   Program Management Team
PPI   Proposal Preparation Instruction
RFP   Request for Proposal
RFQ   Request for Quotation
SCA   Subcontracts Administrator
SCCS   Schedule and Cost control System
SCM   Subcontracts Manager
SCMT   Subcontract Management Team
SDR   System Design Review
SOW   Statement of Work
SRR   System Requirements Review
SSR   Software Specification Review

SWBS   Subcontractor WBS
Ts&Cs Terms and Conditions
TPM   Technical Performance Measurement
TRR   Test Readiness Review
WBS   Work Breakdown Structure

Index

Acquisition process, manufacturing planning and, 7–8
Activity-based costing (ABC), 49
Air Force Acquisition Logistics Division (ALD), 115–118
Army Industrial Productivity Initiative (IPI), 51
AT&T:
  MOS V qualified manufacturing line, 104–113
  Process Quality Management and Improvement (PQMI), 80
  supplier alliance, 209–210
  supplier-management program, 207–209
Audits, 266–267

Bell Helicopter, 102–104
  strategic objectives, 102–103
  tool planning strategy, 103
  trade study process, 104
Benchmarking, competitive, 42–43
Best Manufacturing Practices (BMP) Program (Dept. of the Navy), 52
Best Practices manual, 4, 6, 136, 137
Bid decision, 53–54
Black Box Flowdown Plan, 61

Capability studies, 68
Cause-and-effect diagrams, 83
CIM (*see* Computer-integrated manufacturing)
Closed-loop defect control program, 192–194
Closed-loop feedback program, 61
Communication, maintaining, 22

Company mission, 12
Company profile, 12
Competitive benchmarking, 42–43
Compliance check, 266–267
Component equivalents, selecting from, 156–157
Component failure rates, 157–158
Composites, evaluating use of, 164–166
Computer-integrated manufacturing (CIM), 71–72, 74
Concurrent engineering philosophy, 65
  implementing, 22–23
Configuration management:
  change control and, 59
  subcontractors and, 284–285
Configuration management plan (CMP), 56
Connectivity, maintaining, 22
Consistency, maintaining, 22
Constraints, assessing, 43–44
Continuous improvement, implementing, 23
Contract close-out, 293
Contract data requirements list (CDRL), 56
Coopers & Lybrand, "Made in America" series, 48
Correspondence, program direction, 290
Cost accounting, assessing, 44–45
Cost and price negotiations, 273
Cost management, use in decision making, 48
Cost of ownership, 168–169
Cost Plus Award Fee (CPAF), 256–257
Cost Plus Fixed Fee (CPFF), 256, 257
Cost Plus Incentive Fee (CPIF), 255–256
Cost reimbursable contracts, 254, 255–257